Data Privacy Games

T0182104

Lei Xu • Chunxiao Jiang • Yi Qian • Yong Ren

Data Privacy Games

 Springer

Lei Xu
School of Computer Science
and Technology
Beijing Institute of Technology
Beijing, China

Yi Qian
Peter Kiewit Institute 206B
University of Nebraska-Lincoln
Omaha, Nebraska, USA

Chunxiao Jiang
Tsinghua Space Center
Tsinghua University
Beijing, China

Yong Ren
Department of Electronic Engineering
Tsinghua University
Beijing, China

ISBN 978-3-030-08586-5   ISBN 978-3-319-77965-2   (eBook)
https://doi.org/10.1007/978-3-319-77965-2

Printed on acid-free paper

This Springer imprint is published by the registered company Springer International Publishing AG part
of Springer Nature.
The registered company address is: Gewerbestrasse 11, 6330 Cham, Switzerland

Preface

With the growing popularity of "big data," the potential value of personal data has attracted more and more attention. Applications built on personal data can create tremendous social and economic benefits. Meanwhile, they bring serious threats to individual privacy. The extensive collection, analysis, and transaction of personal data make it difficult for an individual to keep the privacy safe. People now show more concerns about privacy than ever before. How to make a balance between the exploitation of personal information and the protection of individual privacy has become an urgent issue.

In this book, we use methodologies from economics, especially game theory, to investigate solutions to the balance issue. We investigate the strategies of stakeholders involved in the use of personal data and try to find the equilibrium. Specifically, we conduct the following studies.

Considering that data mining is the core technology of big data, in Chap. 1 we propose a user role-based methodology to investigate the privacy issues in data mining. We identify four different types of users, i.e., four *user roles*, involved in data mining applications, including *data provider*, *data collector*, *data miner*, and *decision maker*. For each user role, we discuss its privacy concerns and the strategies that it can adopt to solve the privacy problems. After clarifying each user role's privacy strategies, we can then analyze the interactions among different user roles.

Among the various approaches that can be applied to privacy issues, we are particularly interested in the game theoretical approach. In Chap. 2, we propose a simple game model to analyze the interactions among data provider, data collector, and data miner. By solving the equilibria of the proposed game, we can get useful guidance on how to deal with the trade-off between privacy and data utility. Then in Chaps. 3 and 4, we elaborate the analysis on data collector's strategies in a setting where the data collector buys data from multiple data providers. In Chap. 3, a contract model is proposed to formulate the behavior rules of the data collector and different data providers. And in Chap. 4, a multi-armed bandit model is proposed to analyze the data collector's pricing strategy.

In Chaps. 5 and 6, we discuss how the owners of data (e.g., an individual or a data miner) deal with the trade-off between privacy and utility in data mining applications. Specifically, in Chap. 5 we study user's rating behavior in collaborative filtering-based recommendation systems. In Chap. 6, we consider a distributed classification scenario where data owners adopt differential privacy techniques to protect privacy. In both of the aforementioned scenarios, each data owner wants to obtain high utility, which is measured by recommendation quality or classification accuracy, without disclosing much privacy. Moreover, different data owners' utilities are correlated. We built game models to formulate the interactions among data owners and propose learning algorithms to find the equilibria.

This book is a collection of our recent research progress on data privacy. Basically, our research follows such a paradigm: for a given application scenario, we first build a model, e.g., a game model, to formalize the interaction among different users, then we find the optimal strategies of users via some learning approach. This book is well suited as a reference book for students and researchers who are interested in the privacy issues. We introduce the necessary concepts in a way that is accessible for readers who don't have a solid background in game theory. We hope that this book can provide the reader with a general understanding of how the economic methodologies can be applied.

The topic of privacy is quite hot in current academia. We can now find many books on this topic. The following two features make this book a unique source for students and researchers.

- This book investigates the privacy-utility trade-off issue by analyzing interactions among multiple stakeholders, which is different from the predominant methodology in current literature.
- This book presents a formalized analysis of the privacy preserving strategies. Specifically, game theory and contract theory, which are widely used in the study of economics, are applied in this book to analyze users' strategies. And reinforcement learning methods are applied to find the optimal strategies.

Writing this book would have been impossible without the help of many people. I would like to express my deepest gratitude to Prof. Qian Yi, Prof. Yong Ren, and Dr. Chunxiao Jiang. And many thanks to Prof. K. J. Ray Liu, Prof. Jianhua Li, Prof. Youjian Zhao, Prof. Jian Yuan, Prof. Jian Wang, Prof. Mohsen Guizani, and Dr. Yan Chen, who have made important contributions to the research work introduced in this book. Last but not least, I thank the people from Springer for their support and encouragement.

Although we made an earnest endeavor for this book, there may still be errors in the book. We would highly appreciate if you contact us when you find any.

Beijing, China   Lei Xu
November 2017

Contents

Chapter 1
The Conflict Between Big Data and Individual Privacy

Abstract With the growing popularity of big data applications, data mining technologies has attracted more and more attention in recent years. In the meantime, the fact that data mining may bring serious threat to individual privacy has become a major concern. How to deal with the conflict between big data and individual privacy is an urgent issue. In this chapter, we review the privacy issues related to data mining in a systematic way, and investigate various approaches that can help to protect privacy. According to the basic procedure of data mining, we identify four different types of users involved in big data applications, namely *data provider*, *data collector*, *data miner* and *decision maker*. For each type of user, we discuss its privacy concerns and the methods it can adopt to protect sensitive information. Basics of related research topics are introduced, and state-of-the-art approaches are reviewed. We also present some preliminary thoughts on future research directions. Specifically, we emphasize the game theoretical approaches that are proposed for analyzing the interactions among different users in a data mining scenario. By differentiating the responsibilities of different users with respect to information security, we'd like to provide some useful insights into the trade-off between data exploration and privacy protection.

1.1 Introduction

As the core technology of big data, data mining has been widely applied in various fields. Data mining is the process of discovering interesting patterns and knowledge from large amounts of data [1]. The term "data mining" is often treated as a synonym for another term "knowledge discovery from data" (KDD) which highlights the goal of the mining process. As shown in Fig. 1.1, during the KDD process, the following steps are performed in an iterative way:

- Step 1: Data preprocessing. Basic operations include data selection (to retrieve data relevant to the KDD task from the database), data cleaning (to remove noise and inconsistent data, to handle the missing data fields, etc.) and data integration (to combine data from multiple sources).

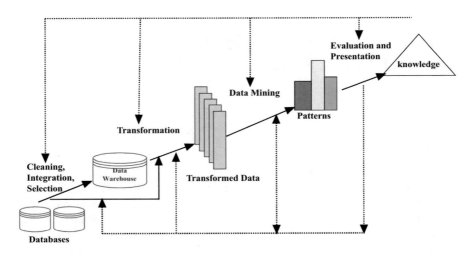

Fig. 1.1 An overview of the KDD process

- Step 2: Data transformation. The goal is to transform data into forms appropriate for the mining task. Basic operations include feature selection and feature transformation.
- Step 3: Data mining. This is an essential process where intelligent methods are employed to extract data patterns (e.g. association rules and classification rules).
- Step 4: Pattern evaluation and presentation. Basic operations include identifying the truly interesting patterns which represent knowledge, and presenting the mined knowledge in an easy-to-understand fashion.

1.1.1 The Privacy Concern and PPDM

The information discovered by data mining can be very valuable, however, people have shown increasing concern about the other side of the coin, namely the privacy threats posed by data mining [2]. Individual privacy may be violated due to the unauthorized access to personal data, the undesired discovery of one's embarrassing information, the use of personal data for purposes other than the one for which data has been collected, etc. There is a conflict between data mining and individual privacy.

To deal with the privacy issues in data mining, a subfield of data mining, referred to as *privacy preserving data mining* (PPDM) has gained a great development in recent years. The objective of PPDM is to safeguard sensitive information from unsolicited or unsanctioned disclosure, and meanwhile, to preserve the utility of data. The consideration of PPDM is twofold. On one hand, sensitive raw data, such as individual's ID card number and cell phone number, should not be directly used for mining. On the other hand, sensitive mining results whose disclosure will result in privacy violation should be excluded. After the pioneering work of Agrawal et al. [3, 4], numerous studies on PPDM have been conducted [5, 6].

1.1.2   User Role-Based Methodology

Current models and algorithms proposed for PPDM mainly focus on how to hide sensitive information from certain mining operations. While, as depicted in Fig. 1.1, KDD is a multi-step process. Besides the mining step, privacy issues may arise in data collecting, data preprocessing, and even in the delivery of the mining results. Considering this, in this chapter we investigate the privacy aspects of data mining by considering the whole knowledge-discovery process. We present an overview of the various approaches that can help to make proper use of sensitive data and protect the sensitive information discovered by data mining [7]. By "sensitive information" we mean the privileged or proprietary information that only certain people are allowed to access. If sensitive information is lost or misused, the subject to which that information belongs will suffer a loss. The term "sensitive data" refers to data from which sensitive information can be extracted. Throughout the book, we use the two terms "privacy" and "sensitive information" interchangeably.

In this chapter, we propose a user-role based methodology to conduct the review of related studies [8]. Based on the stage division of the KDD process, we identify four different types of users, namely four *user roles*, in a typical data mining scenario:

- **Data Provider**: the user who owns some data that are desired by the data mining task.
- **Data Collector**: the user who collects data from data providers and then publishes the data to the data miner.
- **Data Miner**: the user who performs data mining tasks on the data.
- **Decision Maker**: the user who makes decisions based on the data mining results in order to achieve certain goals.

In the scenario depicted in Fig. 1.2, a user represents either a person or an organization. Also, one user can play multiple roles at once.

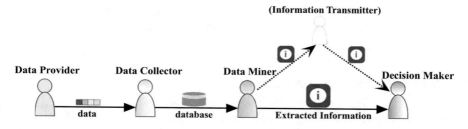

Fig. 1.2 A simple illustration of the application scenario with data mining at the core

By differentiating the four different user roles, we can explore the privacy issues in data mining in a principled way. All users care about the security of sensitive information, but each user role views the security issue from its own perspective. Here we briefly describe the privacy concerns of each user role. Detailed discussions will be presented in following sections.

Data Provider The major concern of a data provider is whether it can control the sensitivity of the data it provides to others. On one hand, the provider should be able to make its private data inaccessible to the data collector. On the other hand, if the provider has to provide some data to the data collector, the provider should get enough compensations for the possible loss in privacy.

Data Collector The data collected from data providers may contain individuals' sensitive information. Directly releasing the data to the data miner will violate data providers' privacy, hence data modification is required. On the other hand, the data should still be useful after modification, otherwise collecting the data will be meaningless. Therefore, the major concern of data collector is to guarantee that the modified data contain no sensitive information but still preserve high utility.

Data Miner The data miner applies mining algorithms to the data provided by data collector. The major concern of data miner is to extract useful information from data in a privacy-preserving manner, i.e. to realize privacy-preserving data mining. As mentioned in Sect. 1.1.1, PPDM covers two types of protections, namely the protection of sensitive data and the protection of sensitive mining results. Here we define that the data collector takes the responsibility of protecting sensitive data, and the data miner mainly focuses on how to hide sensitive mining results from untrustworthy parties.

Decision Maker As shown in Fig. 1.2, a decision maker can get the data mining results directly from the data miner, or from some *information transmitter*. In the latter case, the mining results may be modified by the information transmitter intentionally or unintentionally, which is harmful to the decision maker. Therefore, the major concern of the decision maker is the integrity of the mining results.

Among the various approaches adopted by each user role, we emphasize a common type of approach, namely the game theoretical approach, that can be applied to many privacy problems. Generally, in the data mining scenario, each user pursues maximum interest in terms of privacy or data utility, and the interests of different users are correlated. Thus, the interactions among different users can be modeled as a game [9]. Via game analysis, we can get useful implications on how each user role should behavior so as to solve its privacy problems.

1.1.3 Chapter Organization

The remainder of this chapter is organized as follows: Sects. 1.2–1.5 discuss the privacy problems and approaches to these problems for data provider, data collector, data miner and decision maker, respectively. Game theoretical approaches proposed for privacy issues are reviewed in Sect. 1.6. Finally, this chapter is concluded in Sect. 1.8.

1.2   Data Provider

1.2.1   Concerns of Data Provider

A data provider owns some data from which valuable information can be extracted. In the scenario depicted in Fig. 1.2, there are actually two types of data providers: one is the data provider who provides data to data collector, and the other is the data collector who provides data to data miner. To differentiate the privacy protecting methods adopted by different user roles, here in this section, we restrict ourselves to the ordinary data provider, i.e. the one who owns a relatively small amount of data which contain sensitive information about the provider itself. If a data provider reveals its data to a data collector, the provider's privacy might be comprised due to the unexpected data breach or exposure of sensitive information. Hence, the privacy concern of a data provider is to control what kind of and how much information other people can obtain from its data. Methods that the data provider can adopt to protect privacy are discussed next.

1.2.2   Approaches to Privacy Protection

1.2.2.1   Limit the Access

A data provider provides data to the collector in an active way or a passive way. By "active" we mean that the data provider voluntarily opts in a survey initiated by the data collector, or fill in some registration forms to create an account in a website. By "passive" we mean that the data, which are generated by the provider's routine activities, are recorded by the data collector, while the data provider may have no awareness of the disclosure of its data. When the data provider provides data actively, the provider can simply skip the information that it considers to be sensitive. If the data are passively provided to the data collector, the data provider can take some measures to limit the collector's access to the sensitive data.

For example, if the data provider is an Internet user who is afraid that his online activities may expose his privacy. Then in order to protect privacy, the user can try to erase the traces of his online activities by emptying browser's cache, deleting cookies, clearing usage records of applications, etc. Besides, the provider can utilize the security tools that are developed for Internet environment, such as anti-tracking browser extensions and anti-virus softwares, to protect his data. With the help of the security tools, the user can limit other's access to his personal data. Though there is no guarantee that one's sensitive data can be completely kept out of the reach of untrustworthy data collectors, making a habit of clearing online traces and using security tools does can help to reduce the risk of privacy disclosure.

1.2.2.2   Trade Privacy for Benefit

In some cases, the data provider needs to make a trade-off between the loss of privacy and the benefits brought by participating in data mining. For example, by analyzing a user's demographic information and browsing history, a shopping website can offer personalized product recommendations to the user. The user's sensitive preference may be disclosed but he can enjoy a better shopping experience. If the data provider is able to predict how much benefit it can get by providing its data, then the provider can rationally decide what kind of and how many sensitive data to provide. Consider the following example: a data collector asks the data provider to provide information about his age, gender, occupation and salary. And the data collector tells the data provider how much he would pay for each data item. If the data provider considers salary to be his sensitive information, then based on the prices offered by the collector, he can choose one of the following actions: (1) not to report his salary, if he thinks the price is too low; (2) to report a fuzzy value of his salary, e.g. "less than 10,000 dollars", if he thinks the price is just acceptable; (3) to report an accurate value of his salary, if he thinks the price is high enough.

From the above example we can see that, both the privacy preference of data provider and the incentives offered by data collector will affect the data provider's decision on his data. The data collector can make profit from the data collected from data providers, and the profit heavily depends on the quantity and quality of the data. Hence, data providers' privacy preferences have great influence on data collector's profit which, in turns, affects the data collector's decision on incentives. In order to obtain satisfying benefits by "selling" his data to the data collector, the data provider needs to consider the effect of his decision on data collector's benefits. In the data-selling scenario, both the seller (i.e. the data provider) and the buyer (i.e. the data collector) want to get more benefits, thus the interaction between data provider and data collector can be formally analyzed by using game theory [10]. We will review the applications of game theory in Sect. 1.6.

1.2.2.3   Provide False Data

As discussed above, a data provider can protect his privacy by limiting others' access to his data. However, a disappointed fact that we have to admit is that no matter how hard they try, Internet users cannot completely prevent the unwanted access to their personal information. Thus, in addition to limiting the access, the data provider can provide false information to those untrustworthy data collectors. For example, when using Internet applications, users can create fake identities to protect their privacy. In 2012, Apple Inc. was assigned a patient called "Techniques to pollute electronic profiling" [12] which can help to protect user's privacy. This patent discloses a method for polluting the information gathered by "network eavesdroppers" via making a false online identity for a principal agent, e.g. a service subscriber. The clone identity automatically carries out numerous online actions which are quite different from a user's true activities. When a network eavesdropper collects the data of a user who is utilizing this method, the eavesdropper will be interfered by the massive data created by the clone identity. Real information about of the user is buried under the manufactured phony information.

1.2.3 Summary

Once the data are handed over to others, there is no guarantee that the data provider's sensitive information will be safe. So it is important for data provider to make sure that his sensitive data are out of reach for anyone untrustworthy at the beginning. In principle, the data provider can realize a perfect protection of his privacy by revealing no sensitive data to others, but this may kill the functionality of data mining. In order to enjoy the benefit brought by data mining, sometimes the data provider has to reveal some privacy. A clever data provider should know how to negotiate with the data collector in order to make every piece of the revealed sensitive information worth.

One problem that needs to be highlighted in future research is how to discover privacy disclosure as early as possible. Studies in computer security and network security have developed various techniques for detecting attacks, intrusions and other types of security threats. However, in the context of data mining, the data provider usually has no awareness of how his data are used. Lacking of ways to monitor the behaviors of data collector and data miner, data providers usually learn from media exposure about the invasion of their privacy. According to an investigation report [13], about 62% of data breach incidents take months or even years to be discovered, and nearly 70% of the breaches are discovered by someone other than the data owners. This depressing statistic implies that we are in urgent need of effective methodologies to warn users about privacy incidents in time.

1.3 Data Collector

1.3.1 Concerns of Data Collector

As shown in Fig. 1.2, a data collector collects data from data providers so as to support the subsequent data mining operations. The original data collected from data providers usually contain sensitive information about individuals. If the data collector doesn't take sufficient precautions before releasing the data to public or

data miners, those sensitive information may be disclosed, even though this is not
the collector's original intention. For example, on October 2, 2006, the U.S. online
movie rental service Netflix[1] released a data set containing movie ratings of 500,000
subscribers to the public for a challenging competition called "the Netflix Prize".
The goal of the competition was to improve the accuracy of personalized movie
recommendations. The released data set was supposed to be privacy-safe, since each
data record only contained a subscriber ID (irrelevant with the subscriber's real
identity), the movie info, the rating, and the date on which the subscriber rated
the movie. However, soon after the release, two researchers [14] from University
of Texas found that with a little bit of auxiliary information about an individual
subscriber, an adversary can easily identify the individual's record (if the record is
present in the data set).

The above example implies that it is necessary for the data collector to modify
the original data before releasing them to others, so that sensitive information about
data providers can neither be found in the modified data nor be inferred by anyone
with malicious intent. Generally, the modification will cause a loss in data utility.
The data collector should make sure that sufficient utility of the data can be retained
after the modification, otherwise collecting the data will be a wasted effort. The data
modification process adopted by data collector, with the goal of preserving privacy
and utility simultaneously, is usually called *privacy preserving data publishing*
(PPDP).

Extensive approaches to PPDP have been proposed in last decade. Fung et
al. have systematically summarized and evaluated different approaches in their
frequently cited survey [15]. In this section, we mainly focus on how PPDP is
realized in two emerging applications, namely social networks and location-based
services. To make our review more self-contained, next we first briefly introduce
some basics of PPDP, and then we review studies on social networks and location-
based services respectively.

1.3.2   Approaches to Privacy Protection

1.3.2.1   Basics of PPDP

PPDP mainly studies anonymization approaches for publishing useful data while
preserving privacy. The original data is assumed to be a private table consisting of
multiple records. Each record consists of the following four types of attributes:

- Identifier (ID): Attributes that can directly and uniquely identify an individual,
  such as name, ID number and mobile number.
- Quasi-identifier (QID): Attributes that can be linked with external data to re-
  identify individual records.

---

[1]https://www.netflix.com.

- Sensitive Attribute (SA): Attributes that an individual wants to conceal, such as disease and salary.
- Non-sensitive Attribute (NSA): Attributes other than ID, QID and SA.

Before being published to others, the table is anonymized, that is, identifiers are removed and quasi-identifiers are modified. As a result, individual's identity and values of sensitive attributes can be hidden from adversaries.

How the data table should be anonymized mainly depends on how much privacy we want to preserve. Different privacy models have been proposed to quantify the preservation of privacy. Based on the attack model which describes the ability of the adversary in terms of identifying a target individual, privacy models can be roughly classified into two categories. The first category considers that the adversary is able to identify the record of a target individual by linking the record to data from other sources. The second category considers that the adversary has enough background knowledge to carry out a *probabilistic attack*. That is, the adversary is able to make a confident inference about whether the target's record exist in the table or which value the target's sensitive attribute would take. Typical privacy models includes k-anonymity, l-diversity, t-closeness, ϵ-differential privacy (for preventing table linkage and probabilistic attack), etc.

Among the many privacy models, k-anonymity [16] and its variants are most widely used. The idea of k-anonymity is to modify the values of quasi-identifiers in original data table, so that every tuple in the anonymized table is indistinguishable from at least $k - 1$ other tuples along the quasi-identifiers. The anonymized table is called a k-anonymous table. Figure 1.3 shows an example of 2-anonymity. Intuitively, if a table satisfies k-anonymity and the adversary only knows the quasi-identifier values of the target individual, then the probability of the target's record being identified by the adversary will not exceed $1/k$.

To make the data table satisfy the requirement of a specified privacy model, one can apply the following anonymization operations [15]:

Age	Sex	Zipcode	Disease
5	Female	12000	HIV
9	Male	14000	dyspepsia
6	Male	18000	dyspepsia
8	Male	19000	bronchitis
12	Female	21000	HIV
15	Female	22000	cancer
17	Female	26000	pneumonia
19	Male	27000	gastritis
21	Female	33000	flu
24	Female	37000	pneumonia

(a)

Age	Sex	Zipcode	Disease
[1, 10]	People	1****	HIV
[1, 10]	People	1****	dyspepsia
[1, 10]	People	1****	dyspepsia
[1, 10]	People	1****	bronchitis
[11, 20]	People	2****	HIV
[11, 20]	People	2****	cancer
[11, 20]	People	2****	pneumonia
[11, 20]	People	2****	gastritis
[21, 60]	People	3****	flu
[21, 60]	People	3****	pneumonia

(b)

Fig. 1.3 An example of 2-anonymity, where QID={$Age, Sex, Zipcode$}. (a) original table (b) 2-anonymous table

- Generalization. This operation replaces some values with a parent value in the taxonomy of an attribute.
- Suppression. This operation replaces some values with a special value (e.g. a asterisk '*'), indicating that the replaced values are not disclosed.
- Anatomization. This operation does not modify the quasi-identifier or the sensitive attribute, but de-associates the relationship between the two.
- Permutation. This operation de-associates the relationship between a quasi-identifier and a numerical sensitive attribute by partitioning a set of data records into groups and shuffling their sensitive values within each group.
- Perturbation. This operation replaces the original data values with some synthetic data values, so that the statistical information computed from the perturbed data does not differ significantly from the statistical information computed from the original data.

The anonymization operations will reduce the utility of data. The reduction of data utility is usually represented by *information loss*: higher information loss means lower utility of the anonymized data. A fundamental problem of PPDP is how to make a trade-off between privacy and utility.

1.3.2.2   Privacy-Preserving Publishing of Social Network Data

Social networks have gained a great development in recent years. Aiming at discovering interesting social patterns, social network analysis becomes more and more important. To support the analysis, the company who runs a social network application sometimes needs to publish its data to a third party. However, even if the identifiers of individuals are removed from the published data, publication of the network data may lead to privacy disclosure. Therefore, the network data need to be properly anonymized before they are published.

A social network is usually modeled as a graph, where the vertex represents an entity and the edge represents the relationship between two entities. PPDP in the context of social networks mainly deals with anonymizing graph data, which is much more challenging than anonymizing relational table data [17]. First, modeling adversary's background knowledge about the network is harder. For relational data tables, a small set of quasi-identifiers are used to define the attack models. While for the graph data, various information, such as attributes of an entity and relationships between different entities, may be utilized by the adversary. Second, measuring the information loss is harder. It is difficult to determine whether the original network and the anonymized network are different in certain properties of the network. Third, devising anonymization methods for graph data is harder. Anonymizing a group of tuples in a relational table does not affect other tuples. However, when modifying a network, changing one vertex or edge may affect the rest of the network.

Different approaches have been proposed to deal with aforementioned challenges. Comprehensive surveys of approaches to on social network data anonymization can be found in [18, 19]. In this chapter, we briefly review some recent studies, with focus on the following three aspects: attack model, privacy model, and data utility.

Attack Model Given the anonymized network data, adversaries usually rely on background knowledge to de-anonymize individuals and learn relationships between de-anonymized individuals. Peng et al. [20] propose an algorithm called *Seed-and-Grow* to identify users from an anonymized social graph. The algorithm first identifies a seed sub-graph which is either planted by an attacker or divulged by collusion of a small group of users, and then grows the seed larger based on the adversary's existing knowledge of users' social relations. Sun et al. [21] introduce a relationship attack model called *mutual friend attack*, which is based on the number of mutual friends of two connected individuals. Figure 1.4 shows an example of the mutual friend attack. The original social network G with vertex identities is shown in Fig. 1.4a, and Fig. 1.4b shows the corresponding anonymized network where all individuals' names are removed. In this network, only Alice and Bob have four mutual friends. If an adversary knows this information, then he can uniquely re-identify the edge (D, E) in Fig. 1.4b is (*Alice, Bob*). In [22], Tai et al. study a type of attack named *degree attack*. The motivation is that each individual in a social network is inclined to associated with not only a vertex identity but also a community identity, and the community identity reflects some sensitive information about the individual. It has been shown that, based on some background knowledge about vertex degree, even if the adversary cannot precisely identify the vertex corresponding to an individual, community information and neighborhood information can still be inferred. For example, the network shown in Fig. 1.5 consists of two communities, and the community identity reveals sensitive information (i.e. disease status) about its members. Suppose that an adversary knows Jhon has five friends, then he can infer that Jhon has AIDS, even though he is not sure which of the two vertices (vertex 2 and vertex 3) in the anonymized network (Fig. 1.5b) corresponds to Jhon. From above discussion we can see that, the graph

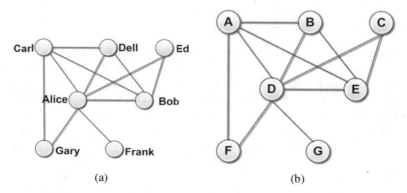

(a)   (b)

Fig. 1.4 Example of mutual friend attack: (**a**) original network; (**b**) naïve anonymized network

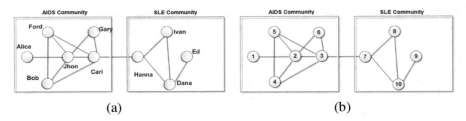

Fig. 1.5 Example of degree attack: (**a**) original network; (**b**) naïve anonymized network

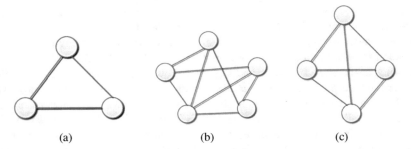

Fig. 1.6 Examples of *k-NMF* anonymity: (**a**) 3-NMF; (**b**) 4-NMF; (**c**) 6-NMF

Fig. 1.7 Examples of
2-structurally diverse graphs,
where the community ID is
indicated beside each vertex:
(**a**) two communities; (**b**)
three communities

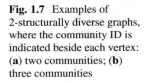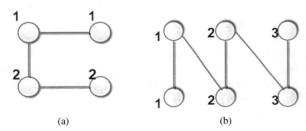

data contain rich information that can be explored by the adversary to initiate an
attack. Modeling the background knowledge of the adversary is difficult yet very
important for deriving the privacy models.

Privacy Model Based on the classic k-anonymity model, a number of privacy
models have been proposed for graph data. Some of the models have been
summarized in the survey [23]. In order to protect the privacy of relationship from
the mutual friend attack, Sun et al. [21] introduce *k-NMF* anonymity. If a network
satisfies *k-NMF* anonymity (see Fig. 1.6), then for each edge e, there will be at least
$k - 1$ other edges with the same number of mutual friends as e. To prevent degree
attacks, Tai et al. [22] introduce the concept of *structural diversity*. A graph satisfies
k-structural diversity anonymization (*k-SDA*), if for every vertex v in the graph,
there are at least k communities, such that each of the communities contains at least
one vertex with the same degree as v (see Fig. 1.7). In other words, for each vertex
v, there are at least $k - 1$ other vertices located in at least $k - 1$ other communities.

Data Utility In the context of network data anonymization, the implication of data utility is: whether and to what extent properties of the graph are preserved. Wu et al. [18] summarize three types of properties considered in current studies. The first type is graph topological properties, which are defined for applications aiming at analyzing graph properties. The second type is graph spectral properties. The third type is aggregate network queries. An aggregate network query calculates the aggregate on some paths or subgraphs satisfying some query conditions. The accuracy of answering aggregate network queries can be considered as the measure of utility preservation. Most existing k-anonymization algorithms for network data publishing perform edge insertion and/or deletion operations, and they try to reduce the utility loss by minimizing the changes on the graph degree sequence.

One important characteristic of social networks is that they evolve over time. Sometimes the data collector needs to publish the network data periodically. The privacy issue in sequential publishing of dynamic social network data has recently attracted much attention. Medforth and Wang [24] identify a new class of privacy attack, named *degree-trail attack*, arising from publishing a sequence of graph data. They demonstrate that even if each published graph is anonymized by strong privacy preserving techniques, an adversary with little background knowledge can re-identify the vertex belonging to a known target individual by comparing the degrees of vertices in the published graphs with the degree evolution of a target. In [25], Tai et al. adopt the same attack model used in [24], and propose a privacy model called dynamic k^w-*structural diversity anonymity* (k^w-*SDA*), for protecting the vertex and multi-community identities in sequential releases of a dynamic network. The parameter k has a similar implication as in the original k-anonymity model, and w denotes a time period that an adversary can monitor a target to collect the attack knowledge. They develop a heuristic algorithm for generating releases satisfying this privacy requirement.

1.3.2.3 Privacy-Preserving Publishing of Trajectory Data

Driven by the increased availability of mobile communication devices with embedded positioning capabilities, location-based services (LBS) have become very popular in recent years. By utilizing the location information of individuals, LBS can bring convenience to our daily life. However, the use of private location information may raise serious privacy problems. Among the many privacy issues in LBS [26, 27], here we focus on the privacy threat brought by publishing trajectory data of individuals. To provide location-based services, commercial entities (e.g. a telecommunication company) and public entities (e.g. a transportation company) collect large amount of individuals' trajectory data, i.e. sequences of consecutive location readings along with time stamps. If the data collector publish such spatio-temporal data to a third party (e.g. a data-mining company), sensitive information about individuals may be disclosed. For example, an advertiser may make inappropriate use of an individual's food preference which is inferred from his frequent visits to some restaurant. To realize a privacy-preserving publication,

id	trajectory
t_1	$a_1 \rightarrow b_1 \rightarrow a_2$
t_2	$a_1 \rightarrow b_1 \rightarrow a_2 \rightarrow b_3$
t_3	$a_1 \rightarrow a_3 \rightarrow b_1$
t_4	$a_3 \rightarrow b_1$
t_5	$a_3 \rightarrow b_2$

(a)

id	trajectory
t_1	$a_1 \rightarrow b_1 \rightarrow a_2$
t_2	$a_1 \rightarrow b_1 \rightarrow a_2$
t_3	$a_3 \rightarrow b_1$
t_4	$a_3 \rightarrow b_1$
t_5	$a_3 \rightarrow b_2$

(b)

Fig. 1.8 Anonymizing trajectory data by suppression [28]. (**a**) original data. (**b**) transformed data

anonymization techniques can be applied to the trajectory data set, so that no sensitive location can be linked to a specific individual. Compared to relational data, spatio-temporal data have some unique characteristics, such as time dependence, location dependence and high dimensionality. Therefore, traditional anonymization approaches cannot be directly applied.

Terrovitis and Mamoulis [28] first investigate the privacy problem in the publication of location sequences. They study how to transform a database of trajectories to a format that would prevent adversaries, who hold a projection of the data, from inferring locations missing in their projections with high certainty. They propose a technique that iteratively suppresses selected locations from the original trajectories until a privacy constraint is satisfied. For example, as shown in Fig. 1.8, if an adversary Jhon knows that his target Mary consecutively visited two location a_1 and a_3, then he can knows for sure that the trajectory t_3 corresponds to Mary, since there is only trajectory that goes through a_1 and a_3. While if some of the locations are suppressed, as shown in Fig. 1.8a, Jhon cannot distinguish between t_3 and t_4, thus the trajectory of Mary is not disclosed. Based on Terrovitis and Mamoulis's work, researchers have now proposed many approaches to solve the privacy problems in trajectory data publishing. Considering that quantification of privacy plays a very important role in the study of PPDP, here we briefly review the privacy models adopted in these studies, especially those proposed in very recent literatures.

Chen et al. [29] assume that, in the context of trajectory data, an adversary's background knowledge on a target individual is bounded by at most L location-time pairs. They propose a privacy model called $(K, C)_L$-privacy for trajectory data anonymization, which considers not only identity linkage attacks on trajectory data, but also attribute linkage attacks via trajectory data. An adversary's background knowledge κ is assumed to be any non-empty subsequence q with $|q| \leq L$ of any trajectory in the trajectory database T. Intuitively, $(K, C)_L$-privacy requires that every subsequence q with $|q| \leq L$ in T is shared by at least a certain number of records,which means the confidence of inferring any sensitive value via q cannot be too high.

Ghasemzadeh et al. [30] propose a method for achieving anonymity in a trajectory database while preserving the information to support effective passenger flow analysis. A privacy model called LK-privacy is adopted in their method to prevent identity linkage attacks. The model assumes that an adversary knows at

id	trajectory
t_1	(d,a,c,e)
t_2	(b,a,e,c)
t_3	(a,d,e)
t_4	(b,d,e,c)
t_5	(d,c)
t_6	(d,e)

(a)

id	trajectory
t_1	$(d,\{a,b,c\},\{a,b,c\},e)$
t_2	$(\{a,b,c\},\{a,b,c\},e,\{a,b,c\})$
t_3	$(\{a,b,c\},d,e)$
t_4	$(\{a,b,c\},d,e,\{a,b,c\})$
t_5	$(d,\{a,b,c\})$
t_6	(d,e)

(b)

Fig. 1.9 Anonymizing trajectory data by generalization [31]. (**a**) original data. (**b**) 2^2-anonymous data

most L previously visited spatio-temporal pairs of any individual. The LK-privacy model requires every subsequence with length at most L in a trajectory database T to be shared by at least K records in T, where L and K are positive integer thresholds. This requirement is quite similar to the $(K, C)_L$-privacy proposed in [29].

Poulis et al. [31] consider previous anonymization methods either produce inaccurate data, or are limited in their privacy specification component. As a result, the cost of data utility is high. To overcome this shortcoming, they propose an approach which applies k^m-anonymity to trajectory data and performs generalization in a way that minimizes the distance between the original trajectory data and the anonymized one. A trajectory is represented by an ordered list of locations that are visited by a moving object. A subtrajectory is formed by removing some locations from the original trajectory, while maintaining the order of the remaining locations. A set of trajectories T satisfies k^m-anonymity if and only if every subtrajectory s of every trajectory $t \in T$, which contains m or fewer locations, is contained in at least k distinct trajectories of T. For example, as shown in Fig. 1.9, if an adversary knows that someone visited location c and then e, then he can infer that the individual corresponds to the trajectory t_1. While given the 2^2-anonymous data, the adversary cannot make a confident inference, since the subtrajectory (c, e) appears in four trajectories.

The privacy models introduced above can all be seen as variants of the classic k-anonymity model. Each model has its own assumptions about the adversary's background knowledge, hence each model has its limitations. A more detailed survey of adversary knowledge, privacy model, and anonymization algorithms proposed for trajectory data publication can be found in [32].

1.3.3 Summary

Privacy-preserving data publishing provides methods to hide identity or sensitive attributes of original data owner. Despite the many advances in the study of data anonymization, there remain some research topics awaiting to be explored. Here we highlight two topics that are important for developing a practically effective anonymization method, namely personalized privacy preservation and modeling the background knowledge of adversaries.

Current studies on PPDP mainly manage to achieve privacy preserving in a statistical sense, that is, they focus on a universal approach that exerts the same amount of preservation for all individuals. While in practice, the implication of privacy varies from person to person. For example, someone considers salary to be sensitive information while someone doesn't; someone cares much about privacy while someone cares less. Therefore, the "personality" of privacy must be taken into account when anonymizing the data. Some researcher have already investigated the issue of personalized privacy preserving. In [33], Xiao and Tao present a generalization framework based on the concept of *personalized anonymity*, where an individual can specify the degree of privacy protection for his sensitive data. Some variants of k-anonymity have also been proposed to support personalized privacy preservation, such as personalized (α, k)-anonymity [34], PK-anonymity [35], individualized (α, k)-anonymity [36], etc. In current studies, individual's personalized preference on privacy preserving is formulated through the parameters of the anonymity model (e.g. the value of k, or the degree of attention paid on certain sensitive value), or nodes in a domain generalization hierarchy. The data provider needs to declare his own privacy requirements when providing data to the collector. However, it is somewhat unrealistic to expect every data provider to define his privacy preference in such a formal way. As "personalization" becomes a trend in current data-driven applications, issues related to personalized data anonymization, such as how to formulate personalized privacy preference in a more flexible way and how to obtain such preference with less effort paid by data providers, need to be further investigated in future research.

The objective of data anonymization is to prevent the potential adversary from discovering information about a certain individual (i.e. the target). The adversary can utilize various kinds of knowledge to dig up the target's information from the published data. From previous discussions on social network data publishing and trajectory data publishing we can see that, if the data collector doesn't have a clear understanding of the capability of the adversary, it is very likely that the anonymized data will be de-anonymized by the adversary. Therefore, in order to design an effective privacy model for preventing various possible attacks, the data collector first needs to make a comprehensive analysis of the adversary's background knowledge and develop proper models to formalize the attacks. However, we are now in an open environment for information exchange, it is difficult to predict from which resources the adversary can retrieve information related to the published data. Besides, as the data type becomes more complex and more advanced data analysis techniques emerge, it is more difficult to determine what kind of knowledge the adversary can learn from the published data. Facing above difficulties, researches should explore more approaches to model adversary's background knowledge. Methodologies from data integration [37], information retrieval, graph data analysis, spatio-temporal data analysis, can be incorporated into this study.

1.4   Data Miner

1.4.1   Concerns of Data Miner

In order to discover useful knowledge which is desired by the decision maker, the data miner applies data mining algorithms to the data obtained from data collector. The privacy issues brought by the data mining operations are twofold. On one hand, if personal information can be directly observed in the data and data breach happens, privacy of the original data owner (i.e. the data provider) will be compromised. On the other hand, equipping with the many powerful data mining techniques, the data miner is able to find out various kinds of information underlying the data. Sometimes the data mining results may reveal sensitive information about the data owners. To encourage data providers to participate in the data mining activity and provide more sensitive data, the data miner needs to make sure that the above privacy threats are eliminated, or in other words, data providers' privacy must be well preserved. As mentioned in Sect. 1.1.2, we consider it is the data collector's responsibility to ensure that sensitive raw data are modified or trimmed out from the published data. The primary concern of data miner is how to prevent sensitive information from appearing in the mining results. To perform a privacy-preserving data mining, the data miner usually needs to modify the data from the data collector. Hence, the decline of data utility is inevitable. Similar to data collector, the data miner also faces the privacy-utility trade-off problem. Specially, quantifications of privacy and utility are closely related to the mining algorithm employed by the data miner.

1.4.2   Approaches to Privacy Protection

Extensive PPDM approaches have been proposed. These approaches can be classified by different criteria [38], such as data distribution, data modification method, data mining algorithm, etc. Based on the distribution of data, PPDM approaches can be classified into two categories, namely approaches for centralized data mining and approaches for distributed data mining. Based on the technique adopted for data modification, PPDM can be classified into perturbation-based, blocking-based, swapping-based, etc. Since we define the privacy goal of data miner as preventing sensitive information from being revealed by the data mining results, in this section, we classify PPDM approaches according to the type of data mining tasks. Specifically, we review recent studies on privacy-preserving association rule mining, privacy-preserving classification, and privacy-preserving clustering, respectively.

1.4.2.1 Privacy-Preserving Association Rule Mining

Association rule mining is one of the most important data mining tasks, which aims at finding interesting associations and correlation relationships among large sets of data items [39]. The problem of mining association rules can be formalized as follows [1]. Given a set of items $I = \{i_1, i_2, \cdots, i_m\}$, and a set of transactions $T = \{t_1, t_2, \cdots, t_n\}$, where each transaction consists of several items from I. An association rule is an implication of the form: $A \Rightarrow B$, where $A \subset I$, $B \subset I$, $A \neq \varnothing$, $B \neq \varnothing$, and $A \cap B \neq \varnothing$. The rule $A \Rightarrow B$ holds in the transaction set T with support s, where s denotes the percentage of transactions in T that contain $A \cup B$. The rule $A \Rightarrow B$ has confidence c in the transaction set T, where c is the percentage of transactions in T containing A that also contain B. Generally, the process of association rule mining contains the following two steps:

- Step 1: Find all frequent itemsets. A set of items is referred to as an *itemset*. The occurrence frequency of an itemset is the number of transactions that contain the itemset. A frequent itemset is an itemset whose occurrence frequency is larger than a predetermined minimum support count.
- Step 2: Generate strong association rules from the frequent itemsets. Rules that satisfy both a minimum support threshold (min_sup) and a minimum confidence threshold (min_conf) are called strong association rules.

Given the thresholds of *support* and *confidence*, the data miner can find a set of association rules from the transactional data set. Some of the rules are considered to be sensitive, either from the data provider's perspective or from the data miner's perspective. To hiding these rules, the data miner can modify the original data set to generate a *sanitized* data set from which sensitive rules cannot be mined, while those non-sensitive ones can still be discovered, at the same thresholds or higher. Various kinds of approaches have been proposed to perform association rule hiding [40], such as heuristic distortion approaches, probabilistic distortion approaches, and reconstruction-based approaches, etc. The main idea behind association rule hiding is to modify the support and/or confidence of certain rules. Here we briefly review some typical modification approaches.

Jain et al. [41] propose a distortion-based approach for hiding sensitive rules, where the position of the sensitive item is altered so that the confidence of the sensitive rule can be reduced, but the support of the sensitive item is never changed and the size of the database remains the same. For example, given the transactional data set shown in Fig. 1.10, set the threshold of support at 33% and the threshold

Fig. 1.10 Altering the position of sensitive item (e.g. *C*) to hide sensitive association rules [41]

Transaction ID	Items	Modified Items
T1	ABC	AB
T2	ABC	ABC
T3	ABC	ABC
T4	AB	AB
T5	A	AC
T6	AC	AC

of confidence at 70%, then the following three rules can be mined from the data: $C \Rightarrow A$ (66.67%, 100%), $A, B \Rightarrow C$ (50%, 75%), $C, A \Rightarrow B$ (50%, 75%). If we consider the item C to be a sensitive item, then we can delete C from the transaction $T1$, and add C to the transaction $T5$. As a result, the above three rules cannot be mined from the modified data set. Dehkoridi [42] considers hiding sensitive rules and keeping the accuracy of transactions as two objectives of some fitness function, and applies genetic algorithm to find the best solution for sanitizing original data. Bonam et al. [43] treat the problem of reducing frequency of sensitive item as a non-linear and multidimensional optimization problem. And they apply *particle swarm optimization* (PSO) technique to this optimization problem.

Among different types of approaches proposed for sensitive rule hiding, we are particularly interested in the reconstruction-based approaches, where a special kind of data mining algorithms, named *inverse frequent set mining* (*IFM*), can be utilized. The problem of IFM was first investigated by Mielikäinen in [44]. The IFM problem can be described as follows [45]: given a collection of frequent itemsets and their support, find a transactional data set such that the data set precisely agrees with the supports of the given frequent itemset collection while the supports of other itemsets would be less than the pre-determined threshold. Guo et al. [46] propose a reconstruction-based approach for association rule hiding where data reconstruction is implemented by solving an IFM problem. As shown in Fig. 1.11, the approach consists of three steps:

- First, use frequent itemset mining algorithm to generate all frequent itemsets with their supports and support counts from original data set.
- Second, determine which itemsets are related to sensitive association rules and remove the sensitive itemsets.
- Third, use the rest itemsets to generate a new transactional data set via inverse frequent set mining.

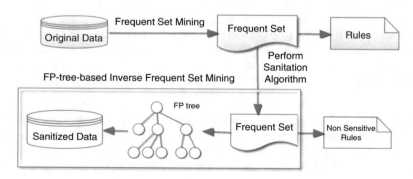

Fig. 1.11 Reconstruction-based association rule hiding [46]

1.4.2.2   Privacy-Preserving Classification

Classification [1] is a form of data analysis that extracts models describing important data classes. Data classification can be seen as a two-step process. In the first step, which is called the *learning* step, a classification algorithm is employed to build a *classifier* (e.g. a classification model) by analyzing a training set made up of tuples and their associated class labels. In the second step, the classifier is used for classification, i.e. predicting categorical class labels of new data. Typical classification model include decision tree, Bayesian model, support vector machine, etc.

Decision Tree A decision tree is a flowchart-like tree structure, where each internal node (non-leaf node) denotes a test on an attribute, each branch represents an outcome of the test, and each leaf node (or terminal node) represents a class label [1]. Given a tuple X, the attribute values of the tuple are tested against the decision tree. A path is traced from the root to a leaf node which holds the class prediction for the tuple. Decision trees can easily be converted to classification rules.

To realize privacy-preserving decision tree mining, Brickell and Shmatikov [47] present a cryptographically secure protocol for privacy-preserving construction of decision trees. The protocol takes place between a user and a server. The user's input consists of the parameters of the decision tree that he wishes to construct, such as which attributes are treated as features and which attribute represents the class. The server's input is a relational database. The user's protocol output is a decision tree constructed from the server's data, while the server learns nothing about the constructed tree. Fong et al. [48] introduce a perturbation and randomization based approach to protect the data sets utilized in decision tree mining. Before being released to a third party for decision tree construction, the original data sets are converted into a group of unreal data sets, from which the original data cannot be reconstructed without the entire group of unreal data sets. Meanwhile, an accurate decision tree can be built directly from the unreal data sets. Sheela and Vijayalakshmi [49] propose a method based on *secure multi-party computation* (SMC) [50] to build a privacy-preserving decision tree over vertically partitioned data. The proposed method utilizes Shamir's secret sharing algorithm to securely compute the cardinality of scalar product, which is needed when computing information gain of attributes during the construction of the decision tree.

Naïve Bayesian Classification Naïve Bayesian classification is based on Bayes' theorem of posterior probability. It assumes that the effect of an attribute value on a given class is independent of the values of other attributes. Given a tuple, a Bayesian classifier can predict the probability that the tuple belongs to a particular class.

Skarkala et al. [51] study the privacy-preserving classification problem for horizontally partitioned data. They propose a privacy-preserving version of the *tree augmented* naïve (TAN) Bayesian classifier [52] to extract global information from horizontally partitioned data. Compared to classical naïve Bayesian classifier, TAN classifier can produce better classification results, since it removes the assumption about conditional independence of attribute. Different from above work, Vaidya et al. [53] consider a centralized scenario, where the data miner has centralized access to a data set. The miner would like to release a classifier on the premise that sensitive information about the original data owners cannot be inferred from the classification model. They utilize differential privacy model [54] to construct a privacy-preserving Naïve Bayesian classifier. The basic idea is to derive the sensitivity for each attribute and to use the sensitivity to compute Laplacian noise. By adding noise to the parameters of the classifier, the data miner can get a classifier which is guaranteed to be differentially private.

Support Vector Machine Support Vector Machine (SVM) is widely used in classification [1]. SVM uses a nonlinear mapping to transform the original training data into a higher dimension. Within this new dimension, SVM searches for a linear optimal separating hyperplane (i.e. a "decision boundary" separating tuples of one class from another), by using *support vectors* and *margins* (defined by the support vectors).

Xia et al. [55] consider that the privacy threat of SVM-based classification comes from the support vectors in the learned classifier. The support vectors are intact instances taken from training data, hence the release of the SVM classifier may disclose sensitive information about the original owner of the training data. They develop a privacy-preserving SVM classifier based on hyperbolic tangent kernel. The kernel function in the classifier is an approximation of the original one. The degree of the approximation, which is determined by the number of support vectors, represents the level of privacy preserving. Lin and Chen [56] also think the release of support vectors will violate individual's privacy. They design a privacy-preserving SVM classifier based on Gaussian kernel function. Privacy-preserving is realized by transforming the original decision function, which is determined by support vectors, to an infinite series of linear combinations of monomial feature mapped support vectors. The sensitive content of support vectors are destroyed by the linear combination, while the decision function can precisely approximate the original one.

1.4.2.3 Privacy-Preserving Clustering

Cluster analysis [1] is the process of grouping a set of data objects into multiple groups or clusters so that objects within a cluster have high similarity, but are very dissimilar to objects in other clusters. Dissimilarities and similarities are assessed based on the attribute values describing the objects and often involve distance measures. Clustering methods can be categorized into partitioning methods, hierarchical methods, density-based methods, etc.

Current studies on privacy-preserving clustering can be roughly categorized into two types, namely approaches based on perturbation and approaches based on secure multi-party computation (SMC).

Perturbation-based approach modifies the data before performing clustering. Oliveira and Zaiane [57] introduce a family of geometric data transformation methods for privacy-preserving clustering. The proposed transformation methods

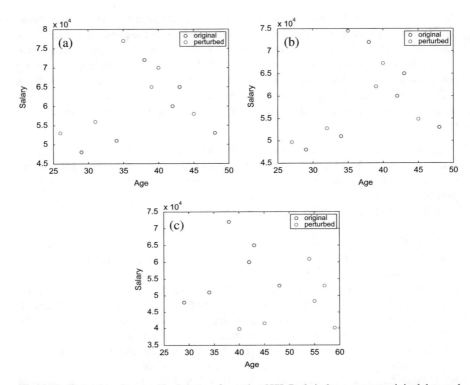

Fig. 1.12 Examples of geometric data transformation [57]. Red circles represent original data and blue circles represent perturbed data. Data are perturbed in three ways: (**a**) translation; (**b**) scaling; (**c**) rotation

distort confidential data attributes by translation, scaling, or rotation (see Fig. 1.12), while general features for cluster analysis are preserved. Oliveira and Zaiane have demonstrated that the transformation methods can well balance privacy and effectiveness, where privacy is evaluated by computing the variance between actual and perturbed values, and effectiveness is evaluated by comparing the number of legitimate points grouped in the original and the distorted databases. The methods proposed in [57] deal with numerical attributes, while in [57], Rajalaxmi and Natarajan propose a set of hybrid data transformations for categorical attributes.

Secure multi-party computation (SMC) is a subfield of cryptography [58]. In general, SMC assumes a number of participants $P_1, P_2, \cdots, P_m$, each has a private data, $X_1, X_2, \cdots, X_m$. The participants want to compute the value of a public function f on m variables at the point $X_1, X_2, \cdots, X_m$. A SMC protocol is called *secure*, if at the end of the computation, no participant knows anything except his own data and the results of global calculation. The SMC-based approaches proposed for clustering make use of primitives from SMC to design a formal model for preserving privacy during the execution of a clustering algorithm. Two pioneer studies on SMC-based clustering are presented in [59, 60]. Vaidya and Clifton [59] present a privacy-preserving method for k-means clustering over vertically partitioned data, where multiple data sites, each having different attributes for the same set of data points, wish to conduct k-means clustering on their joint data. At each iteration of the clustering process, each site can securely find the cluster with the minimum distance for each point, and can independently compute the components of the cluster means corresponding to its attributes. A *checkThreshold* algorithm is proposed to determine whether the stopping criterion is met. Jha et al. [60] design a privacy-preserving k-means clustering algorithm for horizontally partitioned data, where only the cluster means at various steps of the algorithm are revealed to the participating parties. They present two protocols for privacy-preserving computation of cluster means. The first protocol is based on oblivious polynomial evaluation and the second one uses homomorphic encryption. Based on above studies, many privacy-preserving approaches have been developed for k-means clustering. Meskine and Bahloul present an overview of these approaches in [61].

Different from previous studies which focus on k-means clustering, De and Tripathy [62] develop a secure algorithm for hierarchical clustering over vertically partitioned data. There are two parties involved in the computation. In the proposed algorithm, each party first computes k clusters on their own private data set. Then, both parties compute the distance between each data point and each of the k cluster centers. The resulting distance matrices along with the randomized cluster centers are exchanged between the two parties. Based on the information provided by the other party, each party can compute the final clustering result.

1.4.3   Summary

For a data miner, the privacy trouble may come from the discovery of sensitive knowledge, the release of the learned model, or the collaboration with other data miners. To fight against different privacy threats, the data miner needs to take different measures:

1. To prevent sensitive information from appearing in the mining results, the data miner can modify the original data via randomization, blocking, geometric transformation, or reconstruction. The modification often has a negative effect on the utility of the data. To make sure that those non-sensitive information can still be mined from the modified data, the data miner needs to make a balance between privacy and utility. The implications of privacy and data utility vary with the characteristics of data and the purpose of the mining task. As data types become more complex and new types of data mining applications emerge, finding appropriate ways to quantify privacy and utility becomes a challenging task, which is of high priority in future study of PPDM.

2. If the data miner needs to release the model learned (e.g. the decision function of a SVM classifier) from the data to others, the data miner should consider the possibility that some attackers may be able to infer sensitive information from the released model. Compared to privacy-preserving data publishing where attack models and corresponding privacy models have been clearly defined, current studies on PPDM pay less attention to the privacy attacks towards the data mining model. For different data mining algorithms, what kind of sensitive information can be inferred from the parameters of the model, what kind of background knowledge can be utilized by the attacker, and how to modify the model built from data to prevent the disclosure of sensitive information, these problems needs to be explored in future study.
3. When participating in a distributed data mining task, the data miner treats all his data as sensitive data, and his objective is to get the correct mining results without reveal his data to other participators. Various SMC-based approaches have been proposed for privacy-preserving distributed data mining. What kind of information can be exchanged between different participators and how to exchange the information are formally defined by a protocol. However, it is no guarantee that every participator will follow the protocol or truthfully share his data. Interactions among different participators need to be further investigated. Considering the selfish nature of the data miner, game theory may be a proper tool for such problems. Some game theoretical approaches have been proposed for distributed data mining. We will discuss these approaches in Sect. 1.6.

1.5   Decision Maker

1.5.1   Concerns of Decision Maker

The ultimate goal of data mining is to provide useful information to the decision maker, so that the decision maker can choose a better way to achieve his objective, such as increasing sales of products or making correct diagnoses of diseases. At a first glance, it seems that the decision maker has no responsibility for protecting privacy, since we usually interpret privacy as sensitive information about the original data owners (i.e. data providers). Generally, the data miner, the data collector and the data provider are considered to be responsible for the safety of privacy. However, if we look at the privacy issue from a wider perspective, we can see that the decision maker also has privacy concerns. The data mining results provided by the data miner are of high importance to the decision maker. If the results are disclosed to someone else, e.g. a competing company, the decision maker may suffer a loss. That is to say, from the perspective of decision maker, the data mining results are sensitive information. On the other hand, if the decision maker does not get the data mining results directly from the data miner, but from someone else which we called *information transmitter*, the decision maker should be skeptical about the credibility of the results, in case that the results have been distorted. Therefore, the privacy concerns of the decision maker are twofold: how to prevent unwanted disclosure of sensitive mining results, and how to evaluate the credibility of the received mining results.

1.5.2 Approaches to Privacy Protection

To prevent unwanted disclosure of sensitive mining results, usually the decision maker has to resort to legal measures. For example, making a contract with the data miner to forbid the miner from disclosing the mining results to a third party. To determine whether the received information can be trusted, the decision maker can utilize methodologies from data provenance, credibility analysis of web information, or other related research fields. In the rest part of this section, we will first briefly review the studies on data provenance and web information credibility, and then present a preliminary discussion about how these studies can help to analyze the credibility of data mining results.

1.5.2.1 Data Provenance

If the decision maker does not get the data mining results directly from the data miner, he would want to know how the results are delivered to him and what kind of modification may have been applied to the results. This is why "provenance" is needed. The term *provenance* originally refers to the chronology of the ownership, custody or location of a historical object. In information science, a piece of data is treated as the historical object, and *data provenance* refers to the information that helps determine the derivation history of the data, starting from the original source [63]. Two kinds of information can be found in the provenance of the data: the ancestral data from which current data evolved, and the transformations applied to ancestral data that helped to produce current data. With such information, people can better understand the data and judge the credibility of the data.

Since 1990s, data provenance has been extensively studied in the fields of databases and workflows. In [63], Simmhan et al. present a taxonomy of data provenance techniques. The following five aspects are used to capture the characteristics of a provenance system:

- Application of provenance. Provenance systems may be constructed to support a number of uses, such as estimate data quality and data reliability, trace the audit trail of data, repeat the derivation of data, etc.
- Subject of provenance. Provenance information can be collected about different resources present in the data processing system and at various levels of detail.
- Representation of provenance. There are mainly two types of methods to represent provenance information, one is annotation and the other is inversion.

The annotation method uses metadata, which comprise of the derivation history of the data, as annotations and descriptions about sources data and processes. The inversion method uses the property by which some derivations can be inverted to find the input data supplied to derive the output data.

- Provenance storage. Provenance can be tightly coupled to the data it describes and located in the same data storage system or even be embedded within the data file. Alternatively, provenance can be stored separately with other metadata or simply by itself.
- Provenance dissemination. A provenance system can use different ways to disseminate the provenance information, such as providing a derivation graph that users can browse and inspect.

As Internet becomes a major platform for information sharing, provenance of Internet information has attracted some attention. Researchers have developed approaches for information provenance in semantic web [64, 65] and social media [66]. Hartig [64] proposes a provenance model that captures both the information about web-based data access and information about the creation of data. In this model, an ontology-based vocabulary is developed to describe the provenance information. Moreau [65] reviews research issues related to tracking provenance in semantic web. Barbier and Liu [66] study the information provenance problem in social media. They model the social network as a directed graph $G(V, E, p)$, where V is the node set and E is the edge set. Each node in the graph represents an entity and each directed edge represents the direction of information propagation. An information propagation probability p is attached to each edge. Based on the model, they define *the information provenance problem* as follows: given a directed graph $G(V, E, p)$, with known terminals $T \subseteq V$, and a positive integer constant $k \in Z^+$, identify the sources $S \subseteq V$, such that $|S| \leq k$, and $U(S, T)$ is maximized. The function $U(S, T)$ estimates the utility of information propagation which starts from the sources S and stops at the terminals T. To solve this provenance problem, one can leverage the unique features of social networks, e.g. user profiles, user interactions, spatial or temporal information, etc. Two approaches are developed to seek the provenance of information. One approach utilizes the network information to directly seek the provenance of information, and the other approach aims at finding the reverse flows of information propagation.

1.5.2.2   Web Information Credibility

Because of the lack of publishing barriers, the low cost of dissemination, and the lax control of quality, credibility of web information has become a serious issue. Tudjman et al. [67] identify the following five criteria that can be employed by Internet users to differentiate false information from the truth:

- Authority: the real author of false information is usually unclear.
- Accuracy: false information dose not contain accurate data or approved facts.
- Objectivity: false information is often prejudicial.

- Currency: for false information, the data about its source, time and place of its origin is incomplete, out of date, or missing.
- Coverage: false information usually contains no effective links to other information online.

In [68], Metzger summarizes the skills that can help users to assess the credibility of online information.

With the rapid growth of online social media, false information breeds more easily and spreads more widely than before, which further increases the difficulty of judging information credibility. Identifying rumors and their sources in microblogging networks has recently become a hot research topic [69, 70]. Current research usually treats rumor identification as a classification problem, thus the following two issues are involved:

- Preparation of training data set. Current studies usually take rumors that have been confirmed by authorities as positive training samples. Considering the huge amount of messages in microblogging networks, such training samples are far from enough to train a good classifier. Building a large benchmark data set of rumors is in urgent need.
- Feature selection. Various kinds of features can be used to characterize the microblogging messages. In current literature, the following three types of features are often used: content-based features, such as word unigram/bigram, part-of-speech unigram/bigram, text length, number of sentiment word (positive/negative), number of URL, and number of hashtag; user-related features, such as registration time, registration location, number of friends, number of followers, and number of messages posted by the user; network features, such as number of comments and number of retweets.

So far, it is still quite difficult to automatically identifying false information on the Internet. It is necessary to incorporate methodologies from multiple disciplines, such as nature language processing, data mining, machine learning, social networking analysis, and information provenance, into the identification procedure.

1.5.3 Summary

Provenance, which describes where the data came from and how the data evolved over time, can help people evaluate the credibility of data. For a decision maker, if he can acquire complete provenance of the data mining results, then he can easily determine whether the mining results are trustworthy. However, in most cases, provenance of the data mining results is not available. If the mining results are not directly delivered to the decision maker, it is very likely that they are propagated in a less controlled environment. A major approach to represent the provenance information is adding annotations to data. While the reality is that the information transmitter has no motivation to make such annotations, especially when he attempts to alter the original mining results for his own interests. In other words, the possible transformation process of the mining results is non-transparent to the decision maker. In order to support provenance of the data mining results, setting up protocols, which explicitly demand the data miner and information transmitters to append provenance annotations to the data they delivered, is quite necessary. Also, standards which define the essential elements of the annotations should be created, so that the decision maker clearly knows how to interpret the provenance. In addition, techniques that help to automatically create the annotations are desired, with the purpose of reducing the cost of recording provenance information. Above issues should be further investigated in future research, not only because they can help the decision maker judge the credibility of data mining results, but also because they may induce constraints on transmitters' behaviors thus reduce the likelihood of distorted mining results.

Besides provenance, studies on identifying false Internet information also can provide some implications for decision makers. Inspired by the study on rumor identification, we consider it is reasonable to formalize the problem of evaluating credibility of data mining results as a classification problem. If the decision maker has accumulated some credible information from past interactions with the data miner or other reliable sources, a classifier, aiming at distinguishing between fake mining results and truthful results, can be built upon these information. Similar to the studies on microblogs, the decision maker needs to delicately choose the features to characterize the data mining results.

1.6 Game Theory in Data Privacy

1.6.1 Game Theory Preliminaries

In above sections, we have discussed the privacy issues related to data provider, data collector, data miner and decision maker, respectively. Here in this section, we focus on the iterations among different users. When participating in a data mining activity, each user has his own consideration about the benefit he may obtain and the cost he has to pay. For example, a company can make profit from the knowledge mined from customers' data, but he may need to pay high price for data containing sensitive information; a customer can get monetary incentives or better services by providing personal data to the company, but meanwhile he has to consider the potential privacy risks. Generally, the user would act in the way that can bring him more benefits, and one user's action may have effect on other users' benefits. Therefore, it is natural to treat the data mining activity as a *game* played by multiple users, and apply game theoretical approaches to analyze the iterations among different users.

Game theory provides a formal approach to model situations where a group of agents have to choose optimum actions considering the mutual effects of other agents' decisions. The essential elements of a game are *players*, *actions*, *payoffs*, and *information* [9]. Players perform actions at designated times in the game. As a result of the performed actions, players receive payoffs. The payoff to each player depends on both the player's action and other players' actions. Information is modelled using the concept of *information set* which represents a player's knowledge about the values of different variables in the game. The outcome of the game is a set of elements picked from the values of actions, payoffs, and other variables after the game is played out. A player is called *rational* if he acts in such a way as to maximize his payoff. A player's strategy is a rule that tells him which action to choose at each instant of the game, given his information set. A strategy profile is an ordered set consisting of one strategy for each of the players in the game. An *equilibrium* is a strategy profile consisting of a best strategy for each of the players in the game. The most important equilibrium concept for the majority of games is *Nash equilibrium*. A strategy profile is a Nash equilibrium if no player has incentive to deviate from his strategy, given that other players do not deviate.

Game theory has been successfully applied to various fields, such as economics, political science, computer science, etc. Researchers have also employed game theory to deal with the privacy issues related to data mining. In following three subsections we will review some representative game theoretical approaches that are proposed for data collection, distributed data mining and data anonymization.

1.6.2   Private Data Collection and Publication

If a data collector wants to collect data from data providers who place high value on their private data, the collector may need to negotiate with the providers about the "price" of the sensitive data and the level of privacy protection. In [71], Adl et al. build a sequential game model to analyze the private data collection process. In the proposed model, a data user, who wants to buy a data set from the data collector, makes a price offer to the collector at the beginning of the game. If the data collector accepts the offer, he then announces some incentives to data providers in order to collect private data from them. Before selling the collected data to the data user, the data collector applies anonymization technique to the data, in order to protect the privacy of data providers at a certain level. Knowing that data will be anonymized, the data user asks for a privacy protection level that facilitates his most preferable balance between data quality and quantity when making his offer. The data collector also announces a specific privacy protection level to data providers. Based on the protection level and incentives offered by data collector, a data provider decides whether to provide his data. In this data collection game, the level of privacy protection has significant influence on each player's action and payoff. Usually, the data collector and data user have different expectations on the protection level. By solving the subgame perfect Nash equilibriums of the proposed game, a consensus on the level of privacy protection can be achieved. In their later work [72], Adl et al. propose a similar game theoretical approach for aggregate query applications. They show that stable combinations of revelation level which defines how specific data are revealed, retention period of the collected data, price of per data item, and the incentives offered to data providers, can be found by solving the game's equilibria. The game analysis can help to set a privacy policy to achieve maximum revenue while respecting data providers' privacy preferences.

1.6.3 Privacy-Preserving Distributed Data Mining

1.6.3.1 SMC-Based Privacy-Preserving Distributed Data Mining

As mentioned in Sect. 1.4.2, secure multi-party computation (SMC) is widely used in privacy preserving distributed data mining. In a SMC scenario, a set of mutually distrustful parties, each with a private input, jointly compute a function over their inputs. Some protocol is established to ensure that each party can only get the computation result and his own data stay private. However, during the execution of the protocol, a party may take one of the following actions in order to get more benefits:

- Semi-honest adversary: one follows the established protocol and correctly performs the computation but attempts to analyze others' private inputs;
- Malicious adversary: one arbitrarily deviates from the established protocol which leads to the failure of computation.
- Collusion: one colludes with several other parties to expose the private input of another party who doesn't participate in the collusion.

Kargupta et al. [73] formalize the SMC problem as a static game with complete information. By analyzing the Nash equilibriums, they find that if nobody is penalized for dishonest behavior, parties tend to collude. They also propose a cheap-talk based protocol to implement a punishment mechanism which can lead to an equilibrium state corresponding to no collusion. Miyaji et al. [74] propose a two-party secure set-intersection protocol in a game theoretic setting. They assume that parties are neither honest nor corrupt but acted only in their own self-interest. They show that the proposed protocol satisfies computational versions of strict Nash equilibrium and stability with respect to trembles. Ge et al. [75] propose a SMC-based algorithm for privacy preserving distributed association rule mining. The algorithm employs Shamir's secret sharing technique to prevent the collusion of parties. In [76], Nanvati and Jinwala model the secret sharing in distributed association rule mining as a repeated game, where a Nash equilibrium is achieved when all parties send their shares and attain a non-collusive behavior. Based on the game model, they develop punishment policies which aim at getting the maximum possible participants involved in the game so that they can get maximum utilities.

1.6.3.2   Recommender System

Personalized recommendation is a typical application of data mining. The recommendation system predicts users' preference by analyzing the item ratings provided by users, thus the user can protect his private preference by falsifying his ratings. However, false ratings will cause a decline of the quality of recommendation. Halkidi et al. [77] employ game theory to address the trade-off between privacy preservation and high-quality recommendation. In the proposed game model, users are treated as players, and the rating data provided to the recommender server are seen as users' strategies. It has been shown that the Nash equilibrium strategy for each user is to declare false rating only for one item, the one that is highly ranked in his private profile and less correlated with items for which he anticipates recommendation. To find the equilibrium strategy, data exchange between users and the recommender server is modeled as an iterative process. At each iteration, by using the ratings provided by other users at previous iteration, each user computes a rating vector that can maximize the preservation of his privacy, with respect to a constraint of the recommendation quality. Then the user declare this rating vector to the recommender server. After several iterations, the process converges to a Nash equilibrium.

1.6.3.3   Linear Regression as a Non-cooperative Game

Ioannidis and Loiseau [78] study the privacy issue in linear regression modeling. They consider a setting where a data analyst collects private data from multiple individuals to build a linear regression model. In order to protect privacy, individuals add noise to their data, which affects the accuracy of the model. In [78], the interactions among individuals are modeled as a non-cooperative game, where each individual selects the variance level of the noise to minimize his cost. The cost relates to both the privacy loss incurred by the release of data and the accuracy of the estimated linear regression model. It is shown that under appropriate assumptions on privacy and estimation costs, there exists a unique pure Nash equilibrium at which each individual's cost is bounded.

1.6.4   Data Anonymization

Chakravarthy et al. [79] present an interesting application of game theory. They propose a k-anonymity method which utilizes coalitional game theory to achieve a proper privacy level, given the threshold for information loss. The proposed method models each tuple in the data table as a player, and computes the payoff to each player according to a concept hierarchy tree of quasi-identifiers. The equivalent class in the anonymous table is formed by establishing a coalition among different tuples based on their payoffs. Given the affordable information loss, the proposed method can automatically find the most feasible value of k, while traditional methods need to fix up the value of k before the anonymization process.

1.6.5   Assumptions of the Game Model

In above discussions we have reviewed the game theoretical approaches to privacy issues in data mining. Most of the proposed approaches adopt the following research paradigm:

- define the elements of the game, namely the players, the actions and the payoffs;
- determine the type of the game: static or dynamic, complete information or incomplete information;
- solve the game to find equilibriums;
- analyze the equilibriums to obtain some implications for practice.

The above paradigm seems to be simple and clear, while problems in real world can be very complicated. Usually we have to make a few assumptions when developing the game model. Unreasonable assumptions or too many assumptions will hurt the applicability of the game model. For example, the game theoretical approach proposed in [77] assumes that there is an iterative process of data exchange between users and the recommender server. To find the best response to other users' strategies, each user is assumed to be able to get a aggregated version of ratings provided by other users for each item, and can calculate the recommendation result by himself. However, in practical recommendation system, it is unlikely that the user would repeatedly modify the ratings he has already reported to the recommender server. Also, there are so many items in the system, it is unrealistic that a user will collect the ratings of all items. Besides, the recommendation algorithm employed by the recommender server is unknown to the user, hence the user cannot calculate the recommendations by himself. With these improper assumptions, the proposed game analysis can hardly provide meaningful guidance to users. Therefore, we think that future study on game theoretical approaches should pay more attention to the *assumptions*. Real-world problem should be formalized in a more realistic way, so that the game theoretical analysis can have more practical implications.

1.6.6   Mechanism Design and Privacy Protection

Mechanism design is a sub-field of microeconomics and game theory. It considers how to implement good system-wide solutions to problems that involve multiple self-interested agents with private information about their preferences for different outcomes [11]. Incorporating mechanism design into the study of privacy protecting has recently attracted some attention. In a nutshell, a mechanism defines the strategies available and the method used to select the final outcome based on agents' strategies. Specifically, consider a group of n agents $\{i\}$, and each agent i has a privately known type $t_i \in T$. A mechanism $M : T^n \rightarrow O$ is a mapping between (reported) types of the n agents, and some outcome space O. The agent's type t_i determines its preferences over different outcomes. The utility that the agent i with type t_i can get from the outcome $o \in O$ is denoted by $u_i(o, t_i)$. Agents are assumed to be rational, that is, agent i prefers outcome o_1 over o_2 when $u_i(o_1, t_i) > u_i(o_2, t_i)$. The mechanism designer designs the rules of a game, so that the agents will participate in the game and the equilibrium strategies of agents can lead to the designer's desired outcome.

Mechanism design is mostly applied to auction design, where an auction mechanism defines how to determine the winning bidder(s) and how much the bidder should pay for the goods. In the context of data mining, the data collector, who often plays the role of data miner as well, acts as the mechanism designer, and data providers are agents with private information. The data collector wants data providers to participate in the data mining activity, i.e. hand over their private data, but the data providers may choose to opt-out because of the privacy concerns. In order to get useful data mining results, the data collector needs to design mechanisms to encourage data providers to opt-in.

1.6.6.1   Mechanisms for Truthful Data Sharing

A mechanism requires agents to report their preferences over the outcomes. Since the preferences are private information and agents are self-interested, it is likely that the agent would report false preferences. In many cases, the mechanism is expected to be *incentive compatible* [11], that is, reporting one's true preferences should bring the agent larger utility than reporting false preferences. Such mechanism is also called *truthful mechanism*.

Researchers have investigated incentive compatible mechanisms for privacy preserving distributed data mining [80, 81]. In distributed data mining, data required for the mining task are collected from multiple parties. Privacy-preserving methods such as secure multi-party computation protocols can guarantee that only the final result is disclosed. However, there is no guarantee that the data provided by participating parties are truthful. If the data mining function is reversible, that is, given two inputs, x and x', and the result $f(x)$, a data provider is able to calculate $f(x')$, then there is a motivation for the provider to provide false data in order to exclusively learn the correct mining result. To encourage truthful data sharing, Nix and Kantarcioglu [80] model the distributed data mining scenario as an incomplete information game and propose two incentive compatible mechanisms. The first mechanism, which is designed for non-cooperative game, is a Vickrey-Clarke-Groves (VCG) mechanism. The VCG mechanism can encourage truthful data sharing for the risk-averse data provider, and can give a close approximation that encourages minimal deviation from the true data for the risk-neutral data provider. The second mechanism, which is designed for cooperative game, is based on the Shapley value. When data providers form multiple coalitions, this mechanism can create incentives for entire groups of providers to truthfully reveal their data. The practical viability of these two mechanisms have been tested on three data mining models, namely naïve Bayesian classification, decision tree classification, and support vector machine classification. In their later work [81], Nix and Kantarciouglu investigate what kind of privacy-preserving data analysis (PPDA) techniques can be implemented in a way that participating parties have the incentive to provide their true private inputs upon engaging in the corresponding SMC protocols. Under the assumption that participating parties prefer to learn the data analysis result correctly and if possible exclusively, the study shows that several important PPDA tasks including privacy-preserving association rule mining, privacy-preserving naïve Bayesian classification and privacy-preserving decision tree classification are incentive driven. Based on Nix and Kantarcioglu's work, Panoui et al. [82] employ the VCG mechanism to achieve privacy preserving collaborative classification. They consider three types of strategies that a data provider can choose: providing true data, providing perturbed data, or providing randomized data. They show that the use of the VCG mechanism can lead to high accuracy of the data mining task, meanwhile data providers are allowed to provide perturbed data, which means privacy of data providers can be preserved.

1.6.6.2   Privacy Auctions

Aiming at providing support for some specific data mining task, the data collector may ask data providers to provide their sensitive data. The data provider will suffer a loss in privacy if he decides to hand over his sensitive data. In order to motivate data providers to participate in the task, the data collector needs to pay monetary incentives to data providers to compensate their privacy loss. Since different data providers assign different values to their privacy, it is natural for data collector to consider buying private data using an auction. In other words, the data provider can sell his privacy at auction. Ghosh and Roth [83] initiate the study of *privacy auction* in a setting where n individuals selling their binary data to a data analyst. Each individual possesses a private bit $b_i \in \{0, 1\}$ representing his sensitive information (e.g. whether the individual has some embarrassing disease), and reports a cost function c_i to the data analyst who wants to estimate the sum of bits $\sum_{i=1}^{n} b_i$. Differential privacy [54] is employed to quantify the privacy cost: $c_i (\varepsilon) = v_i \cdot \varepsilon$, where v_i is a privately known parameter representing individual's value for privacy, and ε is the parameter of differential privacy. The cost function determines the individual's privacy loss when his private bit b_i is used in an ε-differentially private manner. The compensation (i.e. payment) that an individual can get from the data analyst is determined by a mechanism which takes the cost parameters $\mathbf{v} = (v_1, \cdots, v_n)$ and a collection of private bit values $\mathbf{b} = (b_1, \cdots, b_n)$ as input. In an attempt to maximize his payment, an individual may misreport his value for privacy (i.e. v_i), thus the data collector needs to design truthful mechanisms that can incentivize individuals to report their true privacy cost. Ghosh and Roth study the mechanism design problem for two models, namely insensitive value model and sensitive value model. Insensitive value model considers the privacy cost only incurred by b_i and ignores the potential loss due to the implicit correlations between v_i and b_i. It is shown that truthful mechanism can be derived to help the data analyst achieve a desired trade-off between the accuracy of the estimate and the cost of payments. While the sensitive value model considers that the reported value for privacy also incurs a cost. The study shows that generally, it is impossible to derive truthful mechanisms that can compensate individuals for their privacy loss resulting from the unknown correlation between the private data b_i and the privacy valuation v_i.

To circumvent the impossibility of sensitive value model, Fleischer and Lyu [84] model the correlation between b_i and v_i by assuming that individual's private bit b_i determines a distribution from a set of accurate and publicly known distributions, and the privacy value v_i is drawn from that distribution. Based on this assumption, they design an approximately optimal truthful mechanisms that can produce accurate estimate and protect privacy of both the data (i.e. b_i) and cost (i.e. v_i), when priors of the aforementioned distributions are known. In [85], Ligett and Roth propose a different mechanism which makes no Bayesian assumptions about the distributions of the cost functions. Instead, they assume that the data analyst can randomly approach an individual and make a take-it-or-leave-it offer composing of the payment and differential privacy parameters. The proposed mechanism consists of two algorithms. The first algorithm makes an offer to an individual and receives a binary participation decision. The second algorithm computes an statistic over the private data provided by participating individuals. Nissim et al. [86] bypass the impossibility by assuming that individuals have monotonic privacy valuations, which captures common contexts where certain values for private data are expected to lead to higher valuations for privacy. They develop mechanisms that can incentivize individuals whose privacy valuations are not too large to report their truthful privacy valuations, and output accurate estimations of the sum of private bits, if there are not too many individuals with too-large privacy valuations. The main idea behind the proposed mechanism is to treat the private bit b_i as 0 for all individuals who value privacy too much.

Above studies explore mechanisms for privacy auctions from the perspective of the "buyer", that is, the data providers report their bids (privacy valuations) to the data analyst and the data analyst determine payments to data providers (see Fig. 1.13a). In [87], Riederer et al. study the mechanisms from the seller's perspective. They consider a setting where online users put up sales of their personal information, and information aggregators place bids to gain access to the corresponding user's information (see Fig. 1.13b). They propose a mechanism called Transactional Privacy (TP) that can help users decide what and how much information the aggregators should obtain. This mechanism is based on auction mechanism called the *exponential mechanism* which has been shown to be truthful and can bring approximate optimal revenue for the seller (users in this case). Riederer et al. show that TP can be efficiently implemented when there is a trusted third party. The third party runs an auction where aggregators bid for user's

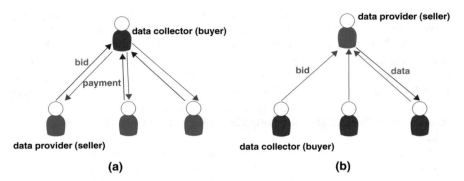

Fig. 1.13 Privacy auction. (**a**) data provider makes a bid (privacy valuation v_i); (**b**) data collector makes a bid (price willing to pay for the data)

information, computes payments to users, and reports to the user about aggregators that received his information. With the proposed mechanism, users can take back control of their personal information.

1.7   Future Research Directions

In previous sections, we have reviewed different approaches to privacy protection for different user roles. Although we have already pointed out some problems that need to be further investigated for each user role, here in this section, we highlight some of the problems and consider them to be the major directions of future research.

1.7.1   Personalized Privacy Preserving

PPDP and PPDM provide methods to explore the utility of data while preserving privacy. However, most current studies only manage to achieve privacy preserving in a statistical sense. Considering that the definition of privacy is essentially personalized, developing methods that can support personalized privacy preserving is an important direction for the study of PPDP and PPDM. As mentioned in Sect. 1.3.3, some researchers have already investigated the issue of personalized anonymization, but most current studies are still in the theoretical stage. Developing practical personalized anonymization methods is in urgent need. Besides, introducing personalized privacy into other types of PPDP/PPDM algorithms is also required. In addition, since complex socioeconomic and psychological factors are involved, quantifying individual's privacy preference is still an open question which expects more exploration.

1.7.2 Data Customization

In Sect. 1.4.2.1 we have discussed that in order to hiding sensitive mining results, we can employ inverse data mining such as inverse frequent set mining to generate data that cannot expose sensitive information. By inverse data mining, we can "customize" the data to get the desired mining result. Alexandra et al. [88] have introduced a concept called *reverse data management* (RDM) which is similar to our specification for inverse data mining. RDM consists of problems where one needs to compute a database input, or modify an existing database input, in order to achieve a desired effect in the output. RDM covers many database problems such as inversion mappings, provenance, data generation, view update, constraint-based repair, etc. We may consider RDM to be a family of data customization methods by which we can get the desired data from which sensitive information cannot be discovered. In a word, data customization can be seen as the inverse process of ordinary data processing. Whenever we have explicit requirements for the outcome of data processing, we may resort to data customization. Exploring ways to solve the inverse problem is an important task for future study.

1.7.3 Provenance for Data Mining

The complete process of data mining consists of multiple phases such as data collection, data preprocessing, data mining, analyzing the extracted information to get knowledge, and applying the knowledge. This process can be seen as an evolvement of data. If the provenance information corresponding to every phase in the process, such as the ownership of data and how the data is processed, can be clearly recorded, it will be much easier to find the origins of security incidents such as sensitive data breach and the distortion of sensitive information. We may say that provenance provides us a way to monitor the process of data mining and the use of mining result. Therefore, techniques and mechanisms that can support provenance in data mining context should receive more attention in future study.

Glavic et al. [89] have discussed how traditional notions of provenance translated to data mining. They identified the need for new types of provenance that can be used to better interpret data mining results. In the context of privacy protection, we are more concerned with how to use provenance to better understand why and how "abnormal" mining result, e.g. result containing sensitive information or false result, appears. Different from provenance approaches that we have reviewed in Sect. 1.5.2.1, approaches for data mining provenance are closely related to the mining algorithm. Therefore, it is necessary to develop new provenance models to specify what kind of provenance information is required and how to present, store, acquire and utilize the provenance information.

1.8   Conclusion

How to protect sensitive information from the security threats brought by data mining has become a hot topic in recent years. In this chapter we review the privacy issues related to data mining by using a user-role based methodology. We differentiate four different user roles that are commonly involved in data mining applications, i.e. data provider, data collector, data miner and decision maker. Each user role has its own privacy concerns, hence the privacy-preserving approaches adopted by one user role are generally different from those adopted by others:

- For data provider, his privacy-preserving objective is to effectively control the amount of sensitive data revealed to others. To achieve this goal, he can utilize security tools to limit other's access to his data, sell his data at auction to get enough compensations for privacy loss, or falsify his data to hide his true identity.
- For data collector, his privacy-preserving objective is to release useful data to data miners without disclosing data providers' identities and sensitive information about them. To achieve this goal, he needs to develop proper privacy models to quantify the possible loss of privacy under different attacks, and apply anonymization techniques to the data.
- For data miner, his privacy-preserving objective is to get correct data mining results while keep sensitive information undisclosed either in the process of data mining or in the mining results. To achieve this goal, he can choose a proper method to modify the data before certain mining algorithms are applied to, or utilize secure computation protocols to ensure the safety of private data and sensitive information contained in the learned model.
- For decision maker, his privacy-preserving objective is to make a correct judgement about the credibility of the data mining results he's got. To achieve this goal, he can utilize provenance techniques to trace back the history of the received information, or build classifier to discriminate true information from false information.

Though different user roles have different privacy concerns, their interests are usually correlated. As we have discussed in Sect. 1.6, game theory is an ideal tool to analyze the interactions among users. Above we have reviewed some game theoretical approaches proposed for privacy issues. In the following chapters, we will introduce some research progress that we've made in this line of work. Specifically, in Chap. 2 we present a simple game model to analyze how the data collector interacts with the data miner and data providers. Then in Chaps. 3 and 4, we focus on the transaction between data collector and data providers. Contract theory and machine learning methods are applied to help the data collector make rational decisions on data price and privacy protection measures. In Chaps. 5 and 6, we discuss how the owners of data (e.g. an individual or a data miner) make decisions on data sharing when participating data mining activities. Different types of game models are established to formalize users' behaviors.

To achieve the privacy-preserving goals of different users roles, various methods from different research fields are required. In this chapter we have reviewed recent progress in related studies, and discussed problems awaiting to be further investigated. We hope that the review presented here can offer researchers different insights into the issue of privacy-preserving data mining, and promote the exploration of new solutions to the security of sensitive information.

References

1. J. Han, M. Kamber, and J. Pei, *Data mining: concepts and techniques.* Morgan Kaufmann, 2006.
2. L. Brankovic and V. Estivill-Castro, "Privacy issues in knowledge discovery and data mining," in *Australian institute of computer ethics conference*, 1999, pp. 89–99.
3. R. Agrawal and R. Srikant, "Privacy-preserving data mining," *SIGMOD Rec.*, vol. 29, no. 2, pp. 439–450, 2000.
4. Y. Lindell and B. Pinkas, "Privacy preserving data mining," in *Advances in Cryptology-CRYPTO 2000.* Springer, 2000, pp. 36–54.
5. C. C. Aggarwal and S. Y. Philip, *A general survey of privacy-preserving data mining models and algorithms.* Springer, 2008.
6. N. Nethravathi, V. J. Desai, P. D. Shenoy, M. Indiramma, and K. Venugopal, "A brief survey on privacy preserving data mining techniques," *Data Mining and Knowledge Engineering*, vol. 8, no. 9, pp. 267–273, 2016.
7. L. Xu, C. Jiang, Y. Chen, J. Wang, and Y. Ren, "A framework for categorizing and applying privacy-preservation techniques in big data mining," *Computer*, vol. 49, no. 2, pp. 54–62, Feb 2016.
8. L. Xu, C. Jiang, J. Wang, J. Yuan, and Y. Ren, "Information security in big data: Privacy and data mining," *IEEE Access*, vol. 2, pp. 1149–1176, 2014.
9. E. Rasmusen and B. Blackwell, *Games and information.* Cambridge, 1994, vol. 2.
10. R. Gibbons, *A primer in game theory.* Harvester Wheatsheaf Hertfordshire, 1992.
11. D. C. Parkes, "Iterative combinatorial auctions: Achieving economic and computational efficiency," Ph.D. dissertation, Philadelphia, PA, USA, 2001.
12. S. Carter, "Techniques to pollute electronic profiling," Apr. 26 2007, US Patent App. 11/257,614. [Online]. Available: https://www.google.com/patents/US20070094738
13. V. C. Inc., "2013 data breach investigations report," 2013. [Online]. Available: http://www.verizonenterprise.com/resources/reports/rp_data-breach-investigations-report-2013_en_xg.pdf
14. A. Narayanan and V. Shmatikov, "How to break anonymity of the netflix prize data set," *The University of Texas at Austin*, 2007.
15. B. Fung, K. Wang, R. Chen, and P. S. Yu, "Privacy-preserving data publishing: A survey of recent developments," *ACM Computing Surveys (CSUR)*, vol. 42, no. 4, p. 14, 2010.
16. L. Sweeney, "k-anonymity: A model for protecting privacy," *International Journal of Uncertainty, Fuzziness and Knowledge-Based Systems*, vol. 10, no. 05, pp. 557–570, 2002.
17. B. Zhou, J. Pei, and W. Luk, "A brief survey on anonymization techniques for privacy preserving publishing of social network data," *ACM SIGKDD Explorations Newsletter*, vol. 10, no. 2, pp. 12–22, 2008.
18. X. Wu, X. Ying, K. Liu, and L. Chen, "A survey of privacy-preservation of graphs and social networks," in *Managing and mining graph data.* Springer, 2010, pp. 421–453.
19. S. Sharma, P. Gupta, and V. Bhatnagar, "Anonymisation in social network: a literature survey and classification," *International Journal of Social Network Mining*, vol. 1, no. 1, pp. 51–66, 2012.

20. W. Peng, F. Li, X. Zou, and J. Wu, "A two-stage deanonymization attack against anonymized social networks," *Computers, IEEE Transactions on*, vol. 63, no. 2, pp. 290–303, Feb 2014.
21. C. Sun, P. S. Yu, X. Kong, and Y. Fu, "Privacy preserving social network publication against mutual friend attacks," *arXiv preprint arXiv:1401.3201*, 2013.
22. C. Tai, P. Yu, D. Yang, and M. Chen, "Structural diversity for resisting community identification in published social networks," 2013.
23. M. Hafez Ninggal and J. Abawajy, "Attack vector analysis and privacy-preserving social network data publishing," in *Trust, Security and Privacy in Computing and Communications (TrustCom), 2011 IEEE 10th International Conference on*. IEEE, 2011, pp. 847–852.
24. N. Medforth and K. Wang, "Privacy risk in graph stream publishing for social network data," in *Data Mining (ICDM), 2011 IEEE 11th International Conference on*. IEEE, 2011, pp. 437–446.
25. C. Tai, P. Tseng, P. Yu, and M. Chen, "Identity protection in sequential releases of dynamic social networks," 2013.
26. G. Ghinita, *Privacy for Location-based Services*, ser. Synthesis Lectures on Information Security, Privacy, and Trust.   Morgan & Claypool Publishers, 2013.
27. M. Wernke, P. Skvortsov, F. Dürr, and K. Rothermel, "A classification of location privacy attacks and approaches," *Personal and Ubiquitous Computing*, vol. 18, no. 1, pp. 163–175, 2014.
28. M. Terrovitis and N. Mamoulis, "Privacy preservation in the publication of trajectories," in *Mobile Data Management, 2008. MDM'08. 9th International Conference on*. IEEE, 2008, pp. 65–72.
29. R. Chen, B. Fung, N. Mohammed, B. C. Desai, and K. Wang, "Privacy-preserving trajectory data publishing by local suppression," *Information Sciences*, vol. 231, pp. 83–97, 2013.
30. M. Ghasemzadeh, B. Fung, R. Chen, and A. Awasthi, "Anonymizing trajectory data for passenger flow analysis," *Transportation Research Part C: Emerging Technologies*, vol. 39, pp. 63–79, 2014.
31. G. Poulis, S. Skiadopoulos, G. Loukides, and A. Gkoulalas-Divanis, "Distance-based kˆ m-anonymization of trajectory data," in *Mobile Data Management (MDM), 2013 IEEE 14th International Conference on*, vol. 2. IEEE, 2013, pp. 57–62.
32. F. Bonchi, L. V. Lakshmanan, and H. W. Wang, "Trajectory anonymity in publishing personal mobility data," *ACM Sigkdd Explorations Newsletter*, vol. 13, no. 1, pp. 30–42, 2011.
33. X. Xiao and Y. Tao, "Personalized privacy preservation," in *Proceedings of the 2006 ACM SIGMOD international conference on Management of data*.   ACM, 2006, pp. 229–240.
34. B. Wang and J. Yang, "Personalized (α, k)-anonymity algorithm based on entropy classification," *Journal of Computational Information Systems*, vol. 8, no. 1, pp. 259–266, 2012.
35. Y. Xua, X. Qina, Z. Yanga, Y. Yanga, and K. Lia, "A personalized k-anonymity privacy preserving method," 2013.
36. S. Yang, L. Lijie, Z. Jianpei, and Y. Jing, "Method for individualized privacy preservation." *International Journal of Security & Its Applications*, vol. 7, no. 6, 2013.
37. A. Halevy, A. Rajaraman, and J. Ordille, "Data integration: the teenage years," in *Proceedings of the 32nd international conference on Very large data bases*. VLDB Endowment, 2006, pp. 9–16.
38. V. S. Verykios, E. Bertino, I. N. Fovino, L. P. Provenza, Y. Saygin, and Y. Theodoridis, "State-of-the-art in privacy preserving data mining," *ACM Sigmod Record*, vol. 33, no. 1, pp. 50–57, 2004.
39. R. Agrawal, T. Imieliński, and A. Swami, "Mining association rules between sets of items in large databases," in *ACM SIGMOD Record*, vol. 22, no. 2. ACM, 1993, pp. 207–216.
40. V. S. Verykios, "Association rule hiding methods," *Wiley Interdisciplinary Reviews: Data Mining and Knowledge Discovery*, vol. 3, no. 1, pp. 28–36, 2013.
41. D. Jain, P. Khatri, R. Soni, and B. K. Chaurasia, "Hiding sensitive association rules without altering the support of sensitive item (s)," in *Advances in Computer Science and Information Technology. Networks and Communications*. Springer, 2012, pp. 500–509.

42. M. N. Dehkordi, "A novel association rule hiding approach in olap data cubes," *Indian Journal of Science & Technology*, vol. 6, no. 2, 2013.
43. J. Bonam, A. R. Reddy, and G. Kalyani, "Privacy preserving in association rule mining by data distortion using pso," in *ICT and Critical Infrastructure: Proceedings of the 48th Annual Convention of Computer Society of India-Vol II*. Springer, 2014, pp. 551–558.
44. T. Mielikäinen, "On inverse frequent set mining," in *Workshop on Privacy Preserving Data Mining*, 2003, pp. 18–23.
45. X. Chen and M. Orlowska, "A further study on inverse frequent set mining," in *Advanced Data Mining and Applications*. Springer, 2005, pp. 753–760.
46. Y. Guo, "Reconstruction-based association rule hiding," in *Proceedings of SIGMOD2007 Ph. D. Workshop on Innovative Database Research*, vol. 2007, 2007, pp. 51–56.
47. J. Brickell and V. Shmatikov, "Privacy-preserving classifier learning," in *Financial Cryptography and Data Security*. Springer, 2009, pp. 128–147.
48. P. K. Fong and J. H. Weber-Jahnke, "Privacy preserving decision tree learning using unrealized data sets," *Knowledge and Data Engineering, IEEE Transactions on*, vol. 24, no. 2, pp. 353–364, 2012.
49. M. A. Sheela and K. Vijayalakshmi, "A novel privacy preserving decision tree induction," in *Information & Communication Technologies (ICT), 2013 IEEE Conference on*. IEEE, 2013, pp. 1075–1079.
50. O. Goldreich, "Secure multi-party computation," *Manuscript. Preliminary version*, 1998. [Online]. Available: http://www.wisdom.weizmann.ac.il/~oded/PS/prot.ps
51. M. E. Skarkala, M. Maragoudakis, S. Gritzalis, and L. Mitrou, "Privacy preserving tree augmented naïve Bayesian multi-party implementation on horizontally partitioned databases," in *Trust, Privacy and Security in Digital Business*. Springer, 2011, pp. 62–73.
52. F. Zheng and G. I. Webb, "Tree augmented naive bayes," in *Encyclopedia of Machine Learning*. Springer, 2010, pp. 990–991.
53. J. Vaidya, B. Shafiq, A. Basu, and Y. Hong, "Differentially private naive bayes classification," in *Web Intelligence (WI) and Intelligent Agent Technologies (IAT), 2013 IEEE/WIC/ACM International Joint Conferences on*, vol. 1. IEEE, 2013, pp. 571–576.
54. C. Dwork, "Differential privacy," in *Automata, languages and programming*. Springer, 2006, pp. 1–12.
55. H. Xia, Y. Fu, J. Zhou, and Y. Fang, "Privacy-preserving svm classifier with hyperbolic tangent kernel," *Journal of Computational Information Systems6*, vol. 5, pp. 1415–1420, 2010.
56. K.-P. Lin and M.-S. Chen, "On the design and analysis of the privacy-preserving svm classifier," *Knowledge and Data Engineering, IEEE Transactions on*, vol. 23, no. 11, pp. 1704–1717, 2011.
57. R. Rajalaxmi and A. Natarajan, "An effective data transformation approach for privacy preserving clustering," *Journal of Computer Science*, vol. 4, no. 4, p. 320, 2008.
58. C. Clifton, M. Kantarcioglu, J. Vaidya, X. Lin, and M. Y. Zhu, "Tools for privacy preserving distributed data mining," *ACM SIGKDD Explorations Newsletter*, vol. 4, no. 2, pp. 28–34, 2002.
59. J. Vaidya and C. Clifton, "Privacy-preserving k-means clustering over vertically partitioned data," in *Proceedings of the ninth ACM SIGKDD international conference on Knowledge discovery and data mining*. ACM, 2003, pp. 206–215.
60. S. Jha, L. Kruger, and P. McDaniel, "Privacy preserving clustering," in *Computer Security–ESORICS 2005*. Springer, 2005, pp. 397–417.
61. R. Akhter, R. J. Chowdhury, K. Emura, T. Islam, M. S. Rahman, and N. Rubaiyat, "Privacy-preserving two-party k-means clustering in malicious model," in *Computer Software and Applications Conference Workshops (COMPSACW), 2013 IEEE 37th Annual*. IEEE, 2013, pp. 121–126.
62. I. De and A. Tripathy, "A secure two party hierarchical clustering approach for vertically partitioned data set with accuracy measure," in *Recent Advances in Intelligent Informatics*. Springer, 2014, pp. 153–162.

63. Y. L. Simmhan, B. Plale, and D. Gannon, "A survey of data provenance in e-science," *ACM Sigmod Record*, vol. 34, no. 3, pp. 31–36, 2005.
64. O. Hartig, "Provenance information in the web of data." in *LDOW*, 2009.
65. L. Moreau, "The foundations for provenance on the web," *Foundations and Trends in Web Science*, vol. 2, no. 2–3, pp. 99–241, 2010.
66. G. Barbier, Z. Feng, P. Gundecha, and H. Liu, "Provenance data in social media," *Synthesis Lectures on Data Mining and Knowledge Discovery*, vol. 4, no. 1, pp. 1–84, 2013.
67. M. Tudjman and N. Mikelic, "Information science: Science about information, misinformation and disinformation," *Proceedings of Informing Science+ Information Technology Education*, pp. 1513–1527, 2003.
68. M. J. Metzger, "Making sense of credibility on the web: Models for evaluating online information and recommendations for future research," *Journal of the American Society for Information Science and Technology*, vol. 58, no. 13, pp. 2078–2091, 2007.
69. F. Yang, Y. Liu, X. Yu, and M. Yang, "Automatic detection of rumor on sina weibo," in *Proceedings of the ACM SIGKDD Workshop on Mining Data Semantics*. ACM, 2012, p. 13.
70. S. Sun, H. Liu, J. He, and X. Du, "Detecting event rumors on sina weibo automatically," in *Web Technologies and Applications*. Springer, 2013, pp. 120–131.
71. R. K. Adl, M. Askari, K. Barker, and R. Safavi-Naini, "Privacy consensus in anonymization systems via game theory," in *Data and Applications Security and Privacy XXVI*. Springer, 2012, pp. 74–89.
72. K. Barker, J. Denzinger, and R. Karimi Adl, "A negotiation game: Establishing stable privacy policies for aggregate reasoning," 2012.
73. H. Kargupta, K. Das, and K. Liu, "Multi-party, privacy-preserving distributed data mining using a game theoretic framework," in *Knowledge Discovery in Databases: PKDD 2007*. Springer, 2007, pp. 523–531.
74. A. Miyaji and M. S. Rahman, "Privacy-preserving data mining: a game-theoretic approach," in *Data and Applications Security and Privacy XXV*. Springer, 2011, pp. 186–200.
75. X. Ge, L. Yan, J. Zhu, and W. Shi, "Privacy-preserving distributed association rule mining based on the secret sharing technique," in *Software Engineering and Data Mining (SEDM), 2010 2nd International Conference on*. IEEE, 2010, pp. 345–350.
76. N. R. Nanavati and D. C. Jinwala, "A novel privacy preserving game theoretic repeated rational secret sharing scheme for distributed data mining," *dcj*, vol. 91, p. 9426611777, 2013.
77. M. Halkidi and I. Koutsopoulos, "A game theoretic framework for data privacy preservation in recommender systems," in *Machine Learning and Knowledge Discovery in Databases*. Springer, 2011, pp. 629–644.
78. S. Ioannidis and P. Loiseau, "Linear regression as a non-cooperative game," in *Web and Internet Economics*. Springer, 2013, pp. 277–290.
79. S. L. Chakravarthy, V. V. Kumari, and C. Sarojini, "A coalitional game theoretic mechanism for privacy preserving publishing based on< i> k</i>-anonymity," *Procedia Technology*, vol. 6, pp. 889–896, 2012.
80. R. Nix and M. Kantarciouglu, "Incentive compatible privacy-preserving distributed classification," *Dependable and Secure Computing, IEEE Transactions on*, vol. 9, no. 4, pp. 451–462, 2012.
81. M. Kantarcioglu and W. Jiang, "Incentive compatible privacy-preserving data analysis," *Knowledge and Data Engineering, IEEE Transactions on*, vol. 25, no. 6, pp. 1323–1335, 2013.
82. A. Panoui, S. Lambotharan, and R. C.-W. Phan, "Vickrey-clarke-groves for privacy-preserving collaborative classification," in *Computer Science and Information Systems (FedCSIS), 2013 Federated Conference on*. IEEE, 2013, pp. 123–128.
83. A. Ghosh and A. Roth, "Selling privacy at auction," in *Proceedings of the 12th ACM conference on Electronic commerce*. ACM, 2011, pp. 199–208.
84. L. K. Fleischer and Y.-H. Lyu, "Approximately optimal auctions for selling privacy when costs are correlated with data," in *Proceedings of the 13th ACM Conference on Electronic Commerce*. ACM, 2012, pp. 568–585.

85. K. Ligett and A. Roth, "Take it or leave it: Running a survey when privacy comes at a cost," in *Internet and Network Economics*.   Springer, 2012, pp. 378–391.
86. K. Nissim, S. Vadhan, and D. Xiao, "Redrawing the boundaries on purchasing data from privacy-sensitive individuals," in *Proceedings of the 5th conference on Innovations in theoretical computer science*.   ACM, 2014, pp. 411–422.
87. C. Riederer, V. Erramilli, A. Chaintreau, B. Krishnamurthy, and P. Rodriguez, "For sale: your data: by: you," in *Proceedings of the 10th ACM WORKSHOP on Hot Topics in Networks*. ACM, 2011, p. 13.
88. A. Meliou, W. Gatterbauer, and D. Suciu, "Reverse data management," *Proceedings of the VLDB Endowment*, vol. 4, no. 12, 2011.
89. B. Glavic, J. Siddique, P. Andritsos, and R. J. Miller, "Provenance for data mining," in *Proceedings of the 5th USENIX Workshop on the Theory and Practice of Provenance*. USENIX Association, 2013, p. 5.

Chapter 2
Privacy-Preserving Data Collecting: A Simple Game Theoretic Approach

Abstract Collecting and publishing personal data may lead to the disclosure of individual privacy. In this chapter, we consider a scenario where a data collector collects data from data providers and then publish the data to a data miner. To protect data providers' privacy, the data collector performs anonymization on the data. Anonymization usually causes a decline of data utility on which the data miner's profit depends, meanwhile, data providers would provide more data if anonymity is strongly guaranteed. How to make a trade-off between privacy protection and data utility is an important question for data collector. We model the interactions among data providers, data collector and data miner as a game. A backward induction-based approach is proposed to find the Nash equilibria of the game. To elaborate the analysis, we also present a specific game formulation which uses k-anonymity as the privacy model. Simulation results show that the game theoretic analysis can help the data collector to achieve a better trade-off between privacy and utility.

2.1 Introduction

As "big data" becomes a hot topic in current days, data mining is attracting more and more attention. Also the value of individuals' data has been widely recognized. In many cases, the data required by a data mining task are collected from individuals by a data collector. The original data may contain sensitive information about individuals. Directly releasing the data to the data miner will cause privacy violation. It is necessary for the data collector to modify the original data before releasing them to others, so that sensitive information about individuals can neither be found in the modified data nor be inferred by anyone with malicious intent. Such a data modification process is usually called *privacy preserving data publishing* (PPDP) [1].

PPDP mainly studies anonymization methods for preserving privacy, while the anonymization operation will reduce the utility of data. How to make a trade-off between privacy and utility is a fundamental issue of PPDP. In this chapter, we study this issue in the following scenario: as depicted in Fig. 2.1, a data miner needs to perform a data mining task, and the miner wants to buy data from a data collector; the data collector collects data for the data miner and applies anonymization

© Springer International Publishing AG, part of Springer Nature 2018
L. Xu et al., *Data Privacy Games*, https://doi.org/10.1007/978-3-319-77965-2_2

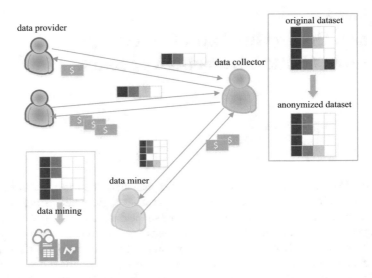

Fig. 2.1 A typical data collecting-publishing-mining scenario

techniques on the collected data; multiple data providers provide data to the data collector, if they think the incentives offered by the collector are appealing and their privacy can be well protected.

The data miner makes profit by mining the data, thus he prefers low degree of anonymization that corresponds to relatively high data utility. On the other hand, data providers prefer high degree of anonymization, so that their privacy can be well protected. When the data collector makes decisions on anonymization, he needs to consider both the data miner's and the data providers' requirements. Also, the degree of anonymization, or we can say the privacy protection level, is affected by how much the data miner will pay to the data collector and whether a data provider will provide his data to the collector. Considering the interactions among different parities, we choose game theory [2] to deal with the data collector's trade-off problem.

Game theory provides a formal approach to model situations where a group of agents have to choose optimum actions considering the mutual effects of other agents' decisions. Several game theoretical approaches have been proposed to deal with the privacy issues in data-driven applications, such as private data collection [3, 5], distributed data mining [6, 7], and personalized recommendation [8]. In [3], Adl et al. proposed a game model to analyze the following data collection process: a data miner, who wants to buy a data set from the data collector, first makes a price offer to the collector. Knowing that the data collector will perform anonymization on the data, the data miner specifies a privacy protection level that facilitates his most preferable balance between data quality and quantity. Then the data collector announces some incentives and a specific privacy protection level to data providers. Usually the data collector and data miner have different expectations on the protection level. By solving the equilibria of the proposed game, a consensus on the level of privacy protection can be achieved.

Inspired by Adl et al.'s work [3], we model the interactions between data miner, data collector and data providers as a sequential game with complete and perfect information [4]. Then we use backward induction to find the game's subgame perfect Nash equilibria [2]. The game model proposed in [3] was developed for deducing a consensual privacy protection level between data miner and data collector. The privacy protection level was set by the data miner. While in our model, it is the data collector who adjusts the privacy protection level to meet the data miner's requirement for data utility.

The rest of the chapter is organized as follows: Sect. 2.2 describes the proposed game model, Sect. 2.3 presents the general approach to find equilibriums. Section 2.4.1 describes a elaborated game formulation by using k-anonymity as the anonymization method. Simulation results are also given in Sect. 2.4.1. This chapter is concluded in Sect. 2.5.

2.2 Game Description

In this section, we describe the basic ingredients of the proposed game.

2.2.1 Players

Players of the game include data providers, data collector and data miner.

Data Providers A data provider decides whether to provide personal data to data collector and how much sensitive information he would like to provide. The decision of a data provider is affected by several factors, including his personal privacy preference (e.g. whether he cares much about privacy), the incentives offered by the data collector, and the level of privacy protection that the data collector guarantees. A data provider's privacy preference is usually unknown to others, which means other players in the game have incomplete information about the data provider. To ease the analysis, instead of considering data providers' decisions individually, we treat all the data providers as a whole. Let D_P denote the data provided by all data providers. The quantity and quality of D_P, together denoted by Q_P, is measured by $Q(D_P)$. The definition of $Q(\cdot)$ can be customized to fit the characters of data and applications. Similar to Adl et al.'s work [3], we assume that, given the incentives and the level of privacy protection offered by data collector, Q_P is deterministic. Then we can focus on the interaction between data collector and data miner. This interaction can be modeled as a sequential game with complete and perfect information.

Data Collector The data collector collects data from data providers and offers incentives to providers. The data collector applies some PPDP technique on the data set D_P, making sure that the level of privacy protection is no less than $\delta_C \in [0, 1]$ that he has promised to data providers. Higher value of δ_C indicates that data providers' privacy will be better protected. The modified data set D_C is then released to the data miner.

Data Miner The data miner buys the data set D_C from data collector and then performs data mining. The profits that the data miner can get largely depend on the quantity and quality of D_C. The data miner is willing to pay higher prices for D_C of higher $Q_C \triangleq Q(D_C)$. To obtain meaningful mining results, the data miner has a minimum requirement for Q_C, denoted by q_M. That is, the data miner won't buy the data D_C if $Q_C < q_M$.

2.2.2 Payoffs

2.2.2.1 Data Miner's Payoff

The data miner makes profit by mining the data D_C. The income of the data miner is defined as:

$$Income_M = g(Q_C),\qquad(2.1)$$

where $g(\cdot)$ is an increasing function of Q_C. The expenditure of data miner is defined as:

$$Expenditure_M = f_M(Q_C; \Theta_M),\qquad(2.2)$$

where $f_M(\cdot\,; \Theta_M)$ is a parametric function. The combination of the parameter(s) Θ_M and the minimum requirement q_M forms an *action* of the data miner. The payoff to the data miner is defined as:

$$G_M = g(Q_C) - f_M(Q_C; \Theta_M).\qquad(2.3)$$

2.2.2.2 Data Collector's Payoff

The income of the data collector is the price paid by the data miner:

$$Income_C = f_M(Q_C; \Theta_M).\qquad(2.4)$$

The expenditure of data collector consists of two parts:

$$Expenditure_C = f_C(Q_P; \Theta_C) + C_C,\qquad(2.5)$$

where $f_C(Q_P; \Theta_C)$ denotes the incentives paid to data providers, C_C denotes a fixed cost of storing and modifying the data. Data collector will pay higher incentives for D_P of higher Q_P. The parametric function $f_C(\cdot\ ; \Theta_C)$ Θ_C is an increasing function of Q_P. The payoff to data collector is defined as:

$$G_C = f_M(Q_C; \Theta_M) - f_C(Q_P; \Theta_C) - C_C. \tag{2.6}$$

The data collector modifies D_P to guarantee the announced privacy protection level δ_C. Higher δ_C means larger decrease in Q_P. We use $T(\delta_C)$ to denote the decrease:

$$Q_C = Q_P - T(\delta_C), \tag{2.7}$$

The combination of Θ_C and δ_C forms an action of the data collector.

The relationship between Q_P and data collector's strategy is formulated as:

$$Q_P = f_P(\Theta_C, \delta_C). \tag{2.8}$$

Plugging (2.7) and (2.8) into (2.6), we can get:

$$\begin{aligned} G_C = & f_M(f_P(\Theta_C, \delta_C) - T(\delta_C); \Theta_M) \\ & - f_C(f_P(\Theta_C, \delta_C); \Theta_C) - C_C. \end{aligned} \tag{2.9}$$

2.2.3 Game Rules

Players choose their actions in sequence. The data miner first makes an offer $O_M = \langle \Theta_M, q_M \rangle$ to data collector. If the data collector rejects this offer, the game terminates and all players get zero payoff. If the offer is accepted, the data collector then announces his offer $O_C = \langle \Theta_C, \delta_C \rangle$ to data providers. Then each data provider makes a response to the offer. The extensive form of this sequential game is shown in Fig. 2.2. Data providers' reaction to each sequence of actions taken by data miner and data collector is represented by the model defined in (2.8), thus data providers are trimmed from the game tree.

2.3 Subgame Perfect Nash Equilibriums

The interaction between data miner and data collector is modeled as a finite sequential game with complete and perfect information. And we can use *backward induction* to find the game's *subgame perfect Nash equilibria* [2].

Fig. 2.2 The trimmed game tree

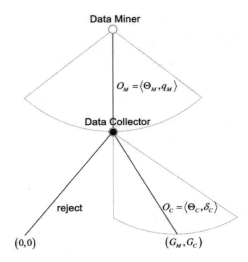

2.3.1   Equilibrium Strategies of Data Collector

According to the principle of backward induction, the first step is to find the optimal actions for data collector in response to each possible action of data miner. Given the offer $O_M = \langle \Theta_M, q_M \rangle$ made by data user, the data collector finds his optimal action by solving the following constrained optimization problem:

$$\max_{\langle \Theta_C, \delta_C \rangle} \; [\, f_M \, (f_P \, (\Theta_C, \delta_C) - T \, (\delta_C) \, ; \, \Theta_M)$$

$$- f_C \, (f_P \, (\Theta_C, \delta_C) \, ; \, \Theta_C) - C_C \,] , \tag{2.10}$$

subject to $0 \leq \delta_C \leq 1$ and $f_P \, (\Theta_C, \delta_C) - T \, (\delta_C) \geq q_M$.

If the optimum solution $O_C^* = \langle \Theta_C^*, \delta_C^* \rangle$ exists and the corresponding maximum payoff G_C^* is greater than zero, then data collector accepts the offer O_M. Otherwise, the data collector rejects the offer. The best response of data collector to a given offer $O_M = \langle \Theta_M, q_M \rangle$ is as follows:

$$BR_C \, (\langle \Theta_M, q_M \rangle) = \begin{cases} Reject, & if \; G_C^* \leq 0 \\ Accept \; with \; \langle \Theta_C^*, \delta_C^* \rangle, & if \; G_C^* > 0 \end{cases} \tag{2.11}$$

2.3.2   Equilibrium Strategies of Data User

The next step to find the equilibria is to determine the optimal action for data miner considering the anticipated reaction of the data collector. When data collector accepts an offer $O_M = \langle \Theta_M, q_M \rangle$ and chooses the action $O_C^* = \langle \Theta_C^*, \delta_C^* \rangle$, the optimal action for data miner can be found by solving the following problem:

$$\underset{\langle \Theta_M, q_M \rangle}{\text{maximize}} \left[\widetilde{g} \left(\Theta_M, q_M \right) - \widetilde{f_M} \left(\Theta_M, q_M \right) \right], \qquad (2.12)$$

where

$$\widetilde{g} \left(\Theta_M, q_M \right) = g \left(f_P \left(\Theta_C^*, \delta_C^* \right) - T \left(\delta_C^* \right) \right),$$

$$\widetilde{f_M} \left(\Theta_M, q_M \right) = f_M \left(f_P \left(\Theta_C^*, \delta_C^* \right) - T \left(\delta_C^* \right) ; \Theta_M \right).$$

We use $O_M^* = \langle \Theta_M^*, q_M^* \rangle$ to denote the optimum solution to above problem.

If both O_M^* and O_C^* exist, the proposed game has two types of subgame perfect Nash equilibria, namely $\left(\langle \Theta_M^*, q_M^* \rangle, \text{Reject} \right)$ and $\left(\langle \Theta_M^*, q_M^* \rangle, \langle \Theta_C^*, \delta_C^* \rangle \right)$, but only the latter has a practical meaning.

2.4 Sample Game Formulation for k-Anonymity

In this section we instantiate the game analysis by specifying the PPDP method adopted by data collector. Here we choose k-anonymity [9] as an example. The basic idea of k-anonymity is to modify the values of quasi-identifiers in original data table, so that every tuple in the anonymized table is indistinguishable from at least $k - 1$ other tuples along the quasi-identifiers.

2.4.1 Game Model

looseness1Suppose that the original data set D_P consists of N tuples. Each tuple corresponds to one data provider and consists of M attributes. Every attribute is allowed to have *null* value. We assume that the total number of potential data providers, denoted by N_0, is a constant. We use D_0 to denote the data set which consists of N_0 tuples and contains no null values. The data set D_P can be seen as the product of replacing some entries of D_0 with null values. As more entries are replaced, the quality of D_P decreases more. We use Q_0 to denote $Q(D_0)$, and define Q_P as follows:

$$Q_P = Q_0 \left(1 - pc_{null} \right), \qquad (2.13)$$

where pc_{null} denotes the percentage of null values in D_P. As defined in (2.8), Q_P depends on data collector's action, which means pc_{null} is determined by Θ_C and

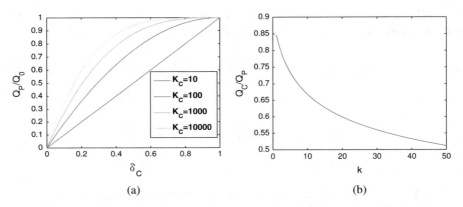

Fig. 2.3 The relationship between data quality and privacy protection. (**a**) Quality of collected data. (**b**) Quality of anonymized data

δ_C. Here for simplicity, we assume the incentive offered by data collector is in proportion to Q_P, that is:

$$f_C(Q_P ; \Theta_C) = K_C Q_P,\qquad(2.14)$$

where $K_C > 1$. Then pc_{null} can be written as:

$$pc_{null} = f_{pc}(K_C, \delta_C).\qquad(2.15)$$

Considering that $0 \le pc_{null} \le 1$ and pc_{null} should decrease as K_C and δ_C increase, we assume $f_{pc}(\cdot)$ has the following form:

$$f_{pc}(K_C, \delta_C) = (1 - \delta_C)^{\log_{10} K_C}.\qquad(2.16)$$

Correspondingly, Q_P can be written as:

$$Q_P = Q_0 \left(1 - (1 - \delta_C)^{\log_{10} K_C}\right).\qquad(2.17)$$

Figure 2.3a shows how Q_P changes with the level of privacy protection. Consider the following two cases: if $\delta_C = 0$, then $Q_P = 0$, because no one wants to provide data if there is no guarantee of privacy security; if $\delta_C = 1$, then $Q_P = Q_0$, which means everyone is willing to provide all required information if their privacy can be fully protected.

Before performing anonymization, the data collector needs to replace the null entries in D_P with appropriate values (e.g. the most common value of the corresponding attribute). The resulting data set is then generalized to D_C which satisfies the criterion of k-anonymity. For a given k ($k \ge 1$), the probability of a tuple in D_C

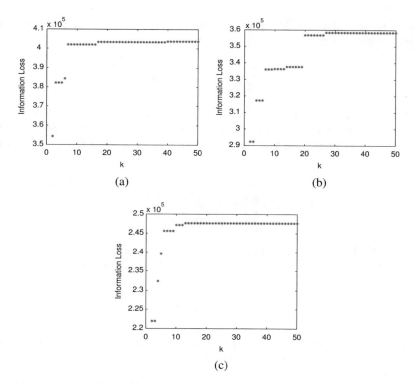

Fig. 2.4 The relationship between information loss and k. We used the k-anonymity algorithm implemented in the tool ARX [13] to perform anonymization on several data sets. The data sets were generated by injecting null values into the original Adult data set [14] which consists of 30,162 tuples and has eight quasi-identifiers. The information loss is measured by non-uniform entropy [12]. (**a**) $pc_{null} = 0$. (**b**) $pc_{null} = 0.15$. (**c**) $pc_{null} = 0.25$

being re-identified is no more than $1/k$. Therefore, we define the privacy protection level δ_C as

$$\delta_C = 1 - \frac{1}{k}. \tag{2.18}$$

The data collector can change the value of k to change the privacy protection level.

The decrease of data utility, or the information loss, can be evaluated by different metrics, such as discernibility [10], classification metric [11], entropy [12], etc. Larger k causes larger information loss, but the specific form of information loss depends on the actual input and output of the anonymization algorithm. By carrying out anonymization experiments on a real data set, we have found that the information loss appears to be a piecewise function of k (see Fig. 2.4). When k is large enough, the information loss is almost invariant with k. To roughly capture the change trend of the information loss, we choose a sigmoid function[1] to model the relationship between k and $(Q_P - Q_C)/Q_P$

$$\frac{Q_P - Q_C}{Q_P} = \frac{k}{\sqrt{k^2 + b}}, \tag{2.19}$$

where b is a constant and $b > 0$. Furthermore, to slow down the increase of information loss with k, we use $\log(k + 2)$ to replace k in above equation. Also we set b to 50 to make the loss range between 0 and 0.5. Then Q_C can be written as:

$$Q_C = Q_P \left(1 - \frac{\log(k + 2)}{\sqrt{(\log(k + 2))^2 + 50}} \right). \tag{2.20}$$

Figure 2.3b shows how Q_C changes with the parameter k.

Similar to the data collector, we assume that the price paid by data miner is in proportion to Q_C:

$$f_M(Q_C; \Theta_M) = K_M Q_C, \tag{2.21}$$

where $K_M > 0$. For a given offer $\langle K_M, q_M \rangle$, the data collector will find the best combination of K_C and k to maximize his payoff:

$$G_C = K_M Q_C - K_C Q_P - C_C. \tag{2.22}$$

The optimal action $\langle K_C^*, k^* \rangle$ should satisfy the following requirement:

$$Q_0 \left(1 - k^{-\log_{10} K_C} \right) \left(1 - \frac{\log(k + 2)}{\sqrt{(\log(k + 2))^2 + 50}} \right) \geq q_M. \tag{2.23}$$

If the maximum payoff G_C^* is larger than zero, then data collector will accept data miner's offer and make an offer $\langle K_C^*, 1 - 1/k^* \rangle$ to data providers. Otherwise, data collector will reject data miner's offer.

The data miner performs data mining on D_C. The income of data miner depends on the specific mining task and the quality of D_C. Here To emphasize the influence of data quality on data miner's income, we define

$$Income_M = g(Q_C) = C_M Q_C, \tag{2.24}$$

where C_M is a constant and $C_M > 0$. Then the payoff to data miner can be defined as:

---

[1]http://en.wikipedia.org/wiki/Sigmoid_function.

Table 2.1 Simulation results

K_M	10	10	10	10	10	10	10	10	10	10
q_M	0	0.1	0.2	0.3	0.4	0.5	0.6	0.7	0.75	0.8
k^*	9	9	9	9	9	10	Fail	Fail	Fail	Fail
K_C^*	3	3	3	3	3	4	Fail	Fail	Fail	Fail
Q_C^*	0.32	0.32	0.32	0.32	0.32	0.35	Fail	Fail	Fail	Fail
G_C^*	2.46	2.46	2.46	2.46	2.46	2.01	Fail	Fail	Fail	Fail
K_M	100	100	100	100	100	100	100	100	100	100
q_M	0	0.1	0.2	0.3	0.4	0.5	0.6	0.7	0.75	0.8
k^*	8	8	8	8	8	8	7	Fail	Fail	Fail
K_C^*	9	9	9	9	9	9	10	Fail	Fail	Fail
Q_C^*	0.43	0.43	0.43	0.43	0.43	0.43	0.44	Fail	Fail	Fail
G_C^*	51.78	51.78	51.78	51.78	51.78	51.78	51.71	Fail	Fail	Fail
K_M	1000	1000	1000	1000	1000	1000	1000	1000	1000	1000
q_M	0	0.1	0.2	0.3	0.4	0.5	0.6	0.7	0.75	0.8
k^*	5	5	5	5	5	5	5	4	Fail	Fail
K_C^*	40	40	40	40	40	40	40	79	Fail	Fail
Q_C^*	0.52	0.52	0.52	0.52	0.52	0.52	0.52	0.55	Fail	Fail
G_C^*	641.95	641.95	641.95	641.95	641.95	641.95	641.95	626.72	Fail	Fail
K_M	10,000	10,000	10,000	10,000	10,000	10,000	10,000	10,000	10,000	10,000
q_M	0	0.1	0.2	0.3	0.4	0.5	0.6	0.7	0.75	0.8
k^*	4	4	4	4	4	4	4	4	3	Fail
K_C^*	191	191	191	191	191	191	191	191	1061	Fail
Q_C^*	0.57	0.57	0.57	0.57	0.57	0.57	0.57	0.57	0.60	Fail
G_C^*	7041.44	7041.44	7041.44	7041.44	7041.44	7041.44	7041.44	7041.44	6477.73	Fail

The value "fail" means that the data collector cannot find any feasible strategy which can both satisfy the data minerŠs requirement and create meaningful profits.

$$G_M = C_M Q_C - K_M Q_C, \qquad (2.25)$$

where Q_C is determined by data collector's strategy $\langle K_C^*, k^* \rangle$, which is actually dependent on K_M and q_M (see (2.22) and (2.23)). Thus the payoff can be rewritten as:

$$G_M = (C_M - K_M) \cdot f_Q (K_M, q_M), \qquad (2.26)$$

where $f_Q (\cdot)$ denotes the relationship between Q_C and $\langle K_M, q_M \rangle$. The data miner searches his optimal action $\langle K_M^*, q_M^* \rangle$ to maximize the above payoff.

2.4.2 Simulation Results

Deriving the analytical forms of K_C^*, k^*, K_M^* and q_M^* is complicated. Here we just choose a group values of K_M and q_M, and conduct simulations in Matlab to find the approximate optimal action $\langle K_C^*, k^* \rangle$ for each combination $\langle K_M, q_M \rangle$. For simplicity, we set $Q_0 = 1$ and $C_C = 0$. Table 2.1 shows the simulation results. The value "fail" means that the data collector cannot find any feasible strategy which can both satisfy the data miner's requirement (see (2.23)) and create meaningful profits (i.e. $G_C > 0$).

Based on the results shown in Table 2.1 we can make following observations:

- As the price parameter K_M increases, the number of "fail" values decreases. For data miner, it means that he can expect the data collector to release data of higher quantity and quality which is beneficial for the mining task.
- As the price parameter K_M increases, the value of k^* decreases. It means that a dishonest data miner would have a greater chance to make extra profits by exploring the privacy information contained in the released data. For data collector, the decrease of k^* means that he will make less efforts to protect data providers' privacy, but instead he has to increase the incentives (larger K_C) to attract the providers, so that he can still collect a data set of desired quantity and quality.
- For a given K_M, the optimum $\langle K_C^*, k^* \rangle$ is almost invariant with q_M. This implicates that under our assumptions about Q_P, Q_C and G_C (see (2.17), (2.20) and (2.22)), the maximum of G_C can be reached at one certain point. Although the increase of q_M will narrow the search space of feasible $\langle K_C, k \rangle$, the maximum point will always be included, as long as q_M is not too high. This phenomenon also suggests that the relationship between Q_P and $\langle K_C, k \rangle$ as well as the quality decrease caused by anonymization need to be further investigated, so that the data miner's requirement q_M can have more influence on data collector's optimal action.

Above observations basically coincide with the intuitions, which shows the validity of the game theoretical analysis. Based on the game analysis results, the data collector can have a general idea about how much effort he needs to pay to protect the privacy of data providers, and the data miner can make a rough estimate of the utility of data and the corresponding expenditure. The data provider will also be more clear about the value of his personal data, thus next time when he provides data to some collector, he will pay more attention to privacy and try to make more profits by making use of his data. In a word, the game theoretical analysis can provide guidance to all parties involved in data mining on how to make a balance between privacy and profit.

2.5   Conclusion

In this chapter, we propose a simple game model to study the interactions among data providers, data collector and data miner. A general approach to find the subgame perfect Nash equilibria of the proposed sequential game is presented. The game theoretical analysis can provide guidance to both the data collector and data miner on the trade-off between data providers' privacy and data utility. In the proposed model, we treat all data providers as a whole. That is, the differences among data providers with respect to privacy preference are ignored. In the following chapters, we will investigate how the data collector interacts with data providers, when data providers are considered individually.

References

1. B. Fung, K. Wang, R. Chen, and P. S. Yu, "Privacy-preserving data publishing: A survey of recent developments," *ACM Comput. Surv.*, vol. 42, no. 4, p. 14, 2010.
2. R. Gibbons, *A primer in game theory.* Harvester Wheatsheaf Hertfordshire, 1992.
3. R. K. Adl, M. Askari, K. Barker, and R. Safavi-Naini, "Privacy consensus in anonymization systems via game theory," in *Data and Applications Security and Privacy XXVI.* Springer, 2012, pp. 74–89.
4. L. Xu, C. Jiang, J. Wang, Y. Ren, J. Yuan, and M. Guizani, "Game theoretic data privacy preservation: Equilibrium and pricing," in *2015 IEEE International Conference on Communications (ICC)*, June 2015, pp. 7071–7076.
5. K. Barker, J. Denzinger, and R. Karimi Adl, "A negotiation game: Establishing stable privacy policies for aggregate reasoning," University of Calgary, Technical Report, 2012. [Online]. Available: http://hdl.handle.net/1880/49282
6. H. Kargupta, K. Das, and K. Liu, "Multi-party, privacy-preserving distributed data mining using a game theoretic framework," in *Knowledge Discovery in Databases: PKDD 2007.* Springer, 2007, pp. 523–531.
7. N. R. Nanavati and D. C. Jinwala, "A novel privacy preserving game theoretic repeated rational secret sharing scheme for distributed data mining," *dcj*, vol. 91, p. 9426611777, 2013.
8. M. Halkidi and I. Koutsopoulos, "A game theoretic framework for data privacy preservation in recommender systems," in *Machine Learning and Knowledge Discovery in Databases.* Springer, 2011, pp. 629–644.
9. L. Sweeney, "k-anonymity: A model for protecting privacy," *International Journal of Uncertainty, Fuzziness and Knowledge-Based Systems*, vol. 10, no. 05, pp. 557–570, 2002.
10. R. J. Bayardo and R. Agrawal, "Data privacy through optimal k-anonymization," in *Data Engineering, 2005. ICDE 2005. Proceedings. 21st International Conference on.* IEEE, 2005, pp. 217–228.
11. V. S. Iyengar, "Transforming data to satisfy privacy constraints," in *Proceedings of the eighth ACM SIGKDD international conference on Knowledge discovery and data mining.* ACM, 2002, pp. 279–288.
12. A. Gionis and T. Tassa, "k-anonymization with minimal loss of information," *Knowledge and Data Engineering, IEEE Transactions on*, vol. 21, no. 2, pp. 206–219, 2009.
13. F. Kohlmayer, F. Prasser, C. Eckert, A. Kemper, and K. Kuhn, "Flash: Efficient, stable and optimal k-anonymity," in *Privacy, Security, Risk and Trust (PASSAT), 2012 International Conference on and 2012 International Confernece on Social Computing (SocialCom)*, 2012, pp. 708–717.
14. K. Bache and M. Lichman, "UCI machine learning repository," 2013. [Online]. Available: http://archive.ics.uci.edu/ml

Chapter 3
Contract-Based Private Data Collecting

Abstract The privacy issues arising in big data applications can be dealt with an economical way. Privacy can be seen as a special type of goods, in a sense that it can be traded by the owner for incentives. In this chapter, we consider a private data collecting scenario where a data collector buys data from multiple data providers and employs anonymization techniques to protect data providers' privacy. Anonymization causes a decline of data utility, therefore, the data provider can only sell his data at a lower price if his privacy is better protected. Achieving a balance between privacy protection and data utility is an important question for the data collector. Considering that different data providers treat privacy differently, and their privacy preferences are unknown to the collector, we propose a contract theoretic approach for data collector to deal with the data providers. By designing an optimal contract, the collector can make rational decisions on how to pay the data providers, and how to protect the providers' privacy. Performance of the proposed contract is evaluated by numerical simulations and experiments on real-world data. The contract analysis shows that when the collector requires a large amount of data, he should ask data providers who care privacy less to provide as much as possible data. We also find that when the collector requires higher utility of data or the data become less profitable, the collector should provide a stronger protection of the providers' privacy.

3.1 Introduction

3.1.1 Data Mining and Privacy Concerns

The success of data mining-based applications requires a sufficient amount of data that may contain private information about individuals. If such data are disclosed or used for purposes other than those initially intended, individual's privacy will be compromised. As we have discussed in Chap. 1, to deal with the privacy issues, substantial work has been done in the field of privacy-preserving data publishing (PPDP) [1] and privacy-preserving data mining (PPDM) [2]. PPDP mainly studies how to *anonymize* data in such a way that after the data is published, individual's

identity and sensitive information cannot be re-identified [3–5]. And PPDM studies
how to prevent sensitive data from being directly used in data mining as well as how
to exclude sensitive mining results [6, 7].

3.1.2 Privacy Auction

Aside from using PPDP and PPDM techniques, the conflict between individual's
demand for privacy safety and commercial application's need for accessing personal
data can be solved in an economic manner [8]. By seeing privacy as a type of goods,
a data collector, who has a need for personal data, can trade with individuals by
paying them compensations. However, since different individuals have different
privacy preferences, e.g. someone cares about privacy very much while someone
cares less, it is difficult for the data collector to decide how to make proper
compensations to different individuals.

A feasible approach to deal with the diversity of individual's privacy preference is
to set up an auction for privacy [9]. At a privacy auction, each individual reports his
valuation on privacy to the data collector. The collector applies some mechanism to
decide how many data he should buy from each individual and how much he should
pay. Ghosh and Roth [10] initiated the study of privacy auction. Based on their work,
a few improved mechanisms have been proposed [11–13]. Current privacy auction
mechanisms are mainly proposed for the *sensitive surveyor's problem* [9], where a
data collector collects individuals' data to obtain an estimate of a simple population
statistic. The private data that an individual owns is represented by a single bit
$b_i \in \{0, 1\}$ indicating whether the individual meets a specified condition, and the
individual's privacy cost is quantified by differential privacy [14]. The objective of
the data collector is to make an accurate estimation of the sum of bits at a low
cost of payments. However, in practice, individual's data is usually represented by
a relational record which consists of multiple attributes. Such representation of data
is the most basic assumption of anonymization algorithms [1]. Therefore, simply
using one bit to represent private data will make the derived auction mechanism less
practical. It is necessary to model the problem with more proper formalizations.

3.1.3 Contract Theoretic Approach

In this chapter, we study the private data collecting problem in a setting where
a data collector collects a set of data records from multiple data providers. Each
data provider gives a certain number of data records to the collector and gets
paid accordingly. To protect data providers' privacy, the data collector applies
anonymization algorithms to the collected data. The anonymized data will then
be used in some data mining task. A high level of anonymization means the

data providers' privacy can be well protected, thus the providers are willing to provide more data or require less compensation. In that sense, anonymization is beneficial to the collector. However, a high level of anonymization also causes a large decrease in data utility, which means the collector will get less benefit from the data. Therefore, the data collector needs to make a trade-off between data utility and privacy protection level. Besides, since different data providers have different privacy preferences, they will react differently to the collector's decision on privacy protection. Considering that the providers' privacy preferences are unknown to the collector, or in other words, there is *information asymmetry* between the collector and providers, it is quite difficult for the collector to make a good trade-off.

Information asymmetry is a common phenomenon in economic activities. For example, when hiring a new employee in the job market, the employer is unable to know exactly the true ability of the employee. As a result, the employer may hire someone who pretends to be capable of the job. A useful tool to deal with the problems caused by information asymmetry is *contract theory* [15]. In the aforementioned example, the employer can sign a contract with the employee to clearly define what kind of work results he expects from the employee and how he will pay the salary. In this chapter, we propose a contract-based approach to handle the trade-off between privacy and utility [16]. Specifically, in the context of private data collection, a contract is signed by the data provider and the data collector to define how many data that the data provider should provide, how much compensation the provider can receive, and to what extent the provider's privacy should be protected. By designing an optimal contract, the data collector can induce the data providers to act in a way that benefits him most.

To solve the optimization problem embedded in the design of optimal contract, we propose a two-step approach which first determines the optimal transfer function for a given level of privacy protection and then optimizes the collector's payoff with respect to the protection level. Due to the complexity of the resulting payoff function, we are unable to explicitly solve the second optimization problem. Instead, based on numerical simulation results, we qualitatively analyze how those external factors, e.g. the data's value to the collector, influence the design of optimal contract. We show such analysis can provide meaningful insight into the data collector's trade-off problem. In addition, by conducting experiments on real data, we have demonstrated that the proposed contract is more beneficial to both the data collector and data providers, when compared to a simple-formed contract which requires the data utility contributed by a data provider to be proportional to how the provider values his privacy.

The rest of this chapter is organized as follows. In Sect. 3.2, we introduce the system model and the contract-theoretic formulation. An elaborative description of the design of optimal contract is presented in Sect. 3.3. In Sect. 3.4, we conduct qualitative analysis of the optimal contract, and evaluate the performance of the two types of contracts through simulations. Finally, we draw conclusions in Sect. 3.5.

3.2   System Model and Problem Formulation

3.2.1   Private Data Collecting

Consider the data collecting scenario shown in Fig. 3.1. A data collector, on the request of some data miner, collects data from N individuals. Each individual, referred to as the data provider, owns a number of data records. The data provider is free to decide how many and what kind of data he would like to provide to the collector. Once handing over his data, the data provider may suffer a loss in privacy. Different data providers may provide same data to the collector. However, when privacy disclosure happens, providers who treat privacy seriously will perceive more loss than those who have little concern about privacy. We use a parameter $\theta \in \left[\underline{\theta}, \bar{\theta}\right]$ ($\underline{\theta} \geq 0$) to describe a data provider's privacy preference. A large θ means the provider cares much about privacy. One thing we do not clarify here is that how the value of the privacy parameter is defined. Quantifying privacy is non-trivial, since complicated sociological and psychological factors may be involved. Here in this chapter, following the conventions of contract theory [15], we think each data provider's θ is decided by the *nature*. The privacy parameter can also be interpreted as the unit cost that the data provider pays for producing data. Let q denote the quantity and quality, together referred to as *utility*, of the data provided by the provider. Then the provider with parameter θ will suffer a monetized loss θq if privacy disclosure happens. Correspondingly, the provider receives a transfer, denoted by t, from the collector as a compensation.

Once the collector has collected enough data, he applies some anonymization technique to the data. After being anonymized, the data becomes more secure, in a sense that the possibility that a data provider is re-identified by an attacker decreases. While in the meantime, the utility of data declines. We use $d(q, \delta)$ to denote the utility of anonymized data, where $\delta \in [0, 1]$ denotes the level of privacy protection that is realized by anonymization. Intuitively, a large δ causes a large decrease in data utility. To embody this intuition, we define $d(q; \delta)$ as

$$d(q, \delta) = \left[\alpha_1(1 - \delta)^{\alpha_2} + \alpha_3\right] q , \qquad (3.1)$$

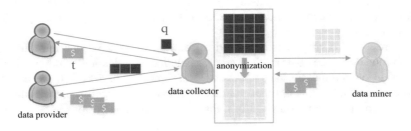

Fig. 3.1 The data collecting scenario

where α_1, α_2 and α_3 are positive constants. This formulation is actually obtained from anonymization experiments on real data (see Sect. 3.4.2.1 for more details). According to the experiment results, there is $0 < \alpha_1 < 1$, $0 < \alpha_2 < 0.5$ and $0 < \alpha_3 < 1$. Here we define $\alpha_3 = 1 - \alpha_1$ to capture the intuition that if no privacy protection measure is taken, i.e. $\delta = 0$, there should be no utility loss.

After finishing the anonymization process, the collector releases the data to a data miner and gets paid, or conducts some analysis by himself. Either way, the collector obtains an income from the data. Let $S(q)$ denote the income, and we assume $S(0) = 0$, $\frac{dS}{dq} > 0$, and $\frac{d^2S}{dq^2} < 0$, which means the marginal value of data decreases as the collector has obtained more data. Furthermore, to ease the analysis and without loss of generality, we define $S(q)$ as

$$S(q) = \lambda \sqrt{q} , \tag{3.2}$$

where the positive constant λ indicates how valuable the data is to the collector. The parameter λ is an exogenous parameter, in a sense that its value is determined by some factors that are out of the control of the data collector and data providers. For example, conditions of the data market will have strong influence on λ. Suppose that a data collector makes profit by selling the collected data to a data miner. If the data miner can buy data from other collectors, then the collector may have to sell the data at a lower price, which implies the data become less valuable to the collector.

Based on above discussions, the payoff to a data provider with parameter θ can be defined as

$$u_\theta = t - (1 - \delta)\theta q , \tag{3.3}$$

where $(1 - \delta)\theta q$ represents the expected value of privacy loss. The payoff that the data collector obtains from the trade with one data provider is

$$u_C = S(d(q, \delta)) - t . \tag{3.4}$$

To maximize the payoff, the data collector needs to carefully decide the transfer paid to the provider and the privacy protection level he should guarantee. However, when trading with a data provider, the collector does not know for sure how the provider values his privacy, since the provider's privacy parameter is only known to himself. In other words, from the perspective of the collector, the privacy parameter θ is a random variable. Here for simplicity and without loss of generality, we make the following assumption.

Assumption 1 The data provider's privacy parameter θ is unknown to the data collector. Each provider's θ is drawn independently and identically from $[\underline{\theta}, \bar{\theta}]$, and the corresponding probability density function $f(\theta)$ is known to the collector.

Realizing that both the data providers and the collector want to get maximal payoff, and there is an information asymmetry between the two parities, we resort to principle-agent theory [15] to solve the collector's problem. More specifically,

Table 3.1 Notations

N	The number of data providers
θ	The privacy preference of a data provider
q	The utility of the data provided by a data provider
t	The transfer paid to a data provider
δ	The level of privacy protection realized by the data collector
β	The probability of privacy disclosure. $\beta = 1 - \delta$
λ	A parameter indicating the data's value to the data collector
q_{req}	The data collector's requirement on the total utility of data
q_{max}	The maximal data utility that a data provider can provide
ρ	The ratio of the data utility provided by a data provider to the maximal data utility that a data provider can provide. $\rho = q/q_{\max}$
$U(\theta)$	The payoff to a data provider with parameter θ
U_C	The payoff to the data collector

we study how to design a contract for the collector, so that the collector can induce data providers to act in a way that can bring him the maximal payoff. Next we will present the formulation of the contract design problem. For convenience, we summarize some important notations used in the formulation in Table 3.1.

3.2.2 Contract-Theoretic Formulation

Following the contract theory terminology, above data collecting scenario can be described as follows. A data collector, who plays the role of the *principal*, delegates a data producing task to multiple *agents*, namely the data providers. Each provider's *type* θ is unobservable to the collector. The collector offers a menu of contracts $\{(\delta, t, q)\}$ to each provider. If the provider chooses to accept the contract (δ, t, q), then he will provide the collector with data of utility q, and in return, the collector should pay transfer t to the provider and make sure that the probability of privacy disclosure is no higher than $1 - \delta$. We assume that the data utility that one data provider can contribute is no more than $q_{\max}$. To make the contract more interpretable, hereafter we use $\rho \triangleq q/q_{\max}$ as a replacement of the contract item q.

According to the revelation principle [15], it is sufficient for the collector to consider only the direct revelation mechanism $\{(\delta(\theta), t(\theta), \rho(\theta))\}$, where the contract $(\delta(\theta), t(\theta), \rho(\theta))$ is designated for data provider with type θ. Considering that most anonymization algorithms do not support personalized privacy protection [1], that is, they exert the same amount of privacy preservation for all individuals, we define $\delta(\theta) = 1 - \beta$ for all $\theta \in [\underline{\theta}, \bar{\theta}]$ with $\beta \in (0, 1]$ denoting the probability of privacy disclosure. Upon choosing the contract $(1 - \beta, t(\theta), \rho(\theta))$, the payoff to a data provider with type θ can be written as

$$U(\theta) = t(\theta) - \beta\theta\rho(\theta) q_{\max} , \tag{3.5}$$

In the study of contract theory, the agent's payoff is usually referred to as *information rent*, which emphasizes that it is because of the information asymmetry that the agent can get extra benefit.

To ensure that the data provider will accept the contract designated for him rather than choosing other contracts or refusing any contract, the menu of contracts must be *incentive feasible*. That is, it should satisfy both the *incentive compatibility* constraints and the *participation* constraints defined below.

Definition 1 A menu of contracts $\{(1 - \beta, t(\theta), \rho(\theta))\}$ is incentive compatible if the best response for the data provider with type θ is to choose the contract $(1 - \beta, t(\theta), \rho(\theta))$ rather than other contracts, i.e., $\forall \left(\theta, \tilde{\theta}\right) \in \left[\underline{\theta}, \bar{\theta}\right]^2$,

$$t(\theta) - \beta\theta\rho(\theta) q_{\max} \geq t\left(\tilde{\theta}\right) - \beta\theta\rho\left(\tilde{\theta}\right) q_{\max} . \tag{3.6}$$

Definition 2 A menu of contracts $\{(1 - \beta, t(\theta), \rho(\theta))\}$ satisfies the participation constraints if it yields to each type of data provider a non-negative payoff, i.e., $\forall \theta \in \left[\underline{\theta}, \bar{\theta}\right]$,

$$t(\theta) - \beta\theta\rho(\theta) q_{\max} \geq 0 . \tag{3.7}$$

In addition, to make sure that meaningful results can be obtained in subsequent data mining tasks, the data collector usually has a minimum requirement on the total utility of the collected data. Here we assume that a feasible menu of contracts should satisfy the following *isoperimetric* constraint:

$$N \int_{\underline{\theta}}^{\bar{\theta}} q_{\max}\rho(\theta) f(\theta) d\theta = q_{req} , \tag{3.8}$$

where q_{req} denotes the data collector's requirement. Apparently, the requirement is attainable only if it is no higher than $N q_{\max}$. Above equation also implies that the total utility of the collected data is assumed to be the summation of the utility of each provider's data. It should be noted that in practice, the relationship between the total utility of data and the utility of each data record is usually application-dependent. Here we define the total utility as a summation, so that we can ease the analysis and meanwhile reflect the general understanding of "total".

Another implicit constraint on the contracts is that the data utility contributed by one data provider is bounded, i.e.

$$0 \leq \rho(\theta) \leq 1 . \tag{3.9}$$

The data collector offers contracts to data providers before knowing the providers' types, hence the payoff that a menu of contracts brings to the collector is evaluated in expected terms. The collector's objective is to find an optimal menu of contracts which satisfies all the constraints listed above and maximizes the expected payoff. The collector's problem can be formulated as

$$\textbf{(P)} \qquad \max_{\{(1-\beta,t(\cdot),\rho(\cdot))\}} N \int_{\underline{\theta}}^{\bar{\theta}} U_C\left(\theta;\beta\right) f\left(\theta\right) d\theta,$$

$$\text{subject to } (3.6)\sim(3.9).$$

The function $U_C\left(\theta;\beta\right)$ in the integrand is defined as

$$U_C\left(\theta;\beta\right) = S\left(d\left(q_{\max}\rho\left(\theta\right), 1-\beta\right)\right) - t\left(\theta\right). \tag{3.10}$$

Next we will discuss how to solve this optimization problem.

3.3   Contract Designs

3.3.1   Method Overview

As defined in the previous section, the contract offered by the collector is formed as a tuple $(1-\beta, t\left(\theta\right), \rho\left(\theta\right))$, where the first item is independent of the provider's type. To find the optimal menu of contracts $\{(1-\beta^*, t^*\left(\theta\right), \rho^*\left(\theta\right))\}$, we propose a two-step approach. First, we find the optimal *transfer function* $t_\beta^*\left(\cdot\right)$ and *production function* $\rho_\beta^*\left(\cdot\right)$ for a given privacy protection level. Specifically, given $\beta \in [0, 1]$, we solve the following problem

$$\textbf{(P1)} \qquad \max_{\{(t(\cdot),\rho(\cdot))\}} \int_{\underline{\theta}}^{\bar{\theta}} U_C\left(\theta;\beta\right) f\left(\theta\right) d\theta ,$$

$$\text{subject to } (3.6)\sim(3.9).$$

Both $t_\beta^*\left(\cdot\right)$ and $\rho_\beta^*\left(\cdot\right)$ can be seen as parametric functions with β being the parameter. By plugging these two functions into the objective function of problem **P**, we can rewrite the data collector's payoff as a function of β, denoted by $U_C\left(\beta\right)$. Thus the second step of optimal contract design is to solve the following optimization problem

$$\textbf{(P2)} \qquad \max_{\beta\in(0,1]} \int_{\underline{\theta}}^{\bar{\theta}} U_C^*\left(\theta;\beta\right) f\left(\theta\right) d\theta .$$

The function $U_C^*\left(\theta;\beta\right)$ in the integrand is defined as

$$U_C^*\left(\theta;\beta\right) = S\left(d\left(q_{\max}\rho_\beta^*\left(\theta\right), 1-\beta\right)\right) - t_\beta^*\left(\theta\right). \tag{3.11}$$

Let β^* denote the optimal solution to above problem, then the optimal menu of contracts is given by $\left\{\left(1-\beta^*, t_{\beta^*}^*\left(\theta\right), \rho_{\beta^*}^*\left(\theta\right)\right)\right\}$.

3.3.2 Simplifying Constraints

Solving problem **P1** is non-trivial, since it involves optimizing a functional with respect to a pair of functions, also the constraints are complicated. Before we explore solutions to the functional optimization problem, we first need to find a concise way to express the incentive constraints and participation constraints.

Though described with one simple inequality, (3.6) actually implies an infinity of constraints, each of which corresponds to a certain pair of θ and $\tilde{\theta}$. Similarly, (3.7) should be treated as an infinity of participation constraints, each of which corresponds to a certain θ. To identify the set of feasible solutions to problem **P1**, first we need to simplify theses constrains as much as possible.

Following a similar approach proposed in [15], we reduce the infinity of incentive constraints in (3.6) to a differential equation

$$\frac{dt\,(\theta)}{d\theta} - \beta q_{\max}\theta \frac{d\rho\,(\theta)}{d\theta} = 0 \tag{3.12}$$

and a monotonicity constraint

$$-\frac{d\rho\,(\theta)}{d\theta} \geq 0 \,. \tag{3.13}$$

Details of the simplification process are presented in the appendix. Further, by using (3.5) we can express (3.12)in a simpler way:

$$\dot{U}\,(\theta) = -\beta q_{\max}\rho\,(\theta) \,. \tag{3.14}$$

Due to the simplicity of above expression, hereafter we focus on the design of $U\,(\cdot)$ instead of $t\,(\cdot)$, after all the optimal $t\,(\cdot)$ can be easily determined once the optimal $U\,(\cdot)$ and $\rho\,(\cdot)$ are found.

Base on (3.9) and (3.14), participation constraints in (3.7) can be simplified to $U\,(\bar{\theta}) \geq 0$. Further, we can predict that this constraint must be binding at the optimum, i.e.

$$U_\beta^*\,(\bar{\theta}) = 0 \,. \tag{3.15}$$

Suppose that $U_\beta^*\,(\bar{\theta}) > 0$, then the collector could reduce $U_\beta^*\,(\bar{\theta})$ by a small amount while keeping $\rho_\beta^*\,(\cdot)$ unchanged. As a result, the collector's payoff is increased, which contradicts with the optimality of $U_\beta^*\,(\cdot)$.

Based on above simplifications, problem **P1** can be rewritten as

$$(\textbf{P1}') \qquad \max_{\{(U(\cdot),\rho(\cdot))\}} \int_{\underline{\theta}}^{\bar{\theta}} U_C\,(\theta;\beta)\,f\,(\theta)\,d\theta \,.$$

subject to (3.13), (3.14), (3.15), (3.8) and (3.9).

The function $U_C(\theta; \beta)$ in the integrand is now written as

$$
\begin{aligned}
U_C(\theta; \beta) =& S\left(d\left(q_{\max}\rho(\theta), 1 - \beta\right)\right) \\
& - U(\theta) - \beta q_{\max}\theta\rho(\theta) \ .
\end{aligned}
\tag{3.16}
$$

3.3.3 Optimal Control-Based Approach

Problem **P1$'$** fits the general formulation of the optimal control problem [17], hence methods developed for optimal control can be applied. Let $y(\theta) \triangleq \rho(\theta)$ be the control variable and $x_1(\theta) \triangleq U(\theta)$ be the state variable. To handle the isoperimetric constraint (3.8), a new state variable $x_2(\theta)$ is defined, and it satisfies the following differential equation:

$$
\dot{x}_2(\theta) = q_{\max} y(\theta) f(\theta) \ ,
\tag{3.17}
$$

The boundary conditions of $x_2(\theta)$ are $x_2(\bar{\theta}) = {}^{q_{req}}/N_{q_{\max}}$. The *Hamiltonian* is

$$
\begin{aligned}
& H\left(\mathbf{x}(\theta), y(\theta), \mathbf{p}(\theta), \theta\right) \\
& = \left[S\left(d\left(q_{\max} y(\theta); 1 - \beta\right)\right) - x_1(\theta) - \beta q_{\max}\theta y(\theta) \right] f(\theta) \\
& - \beta q_{\max} p_1(\theta) y(\theta) + p_2(\theta) q_{\max} f(\theta) y(\theta) \ ,
\end{aligned}
\tag{3.18}
$$

where $p_1(\theta)$ and $p_1(\theta)$ are co-state variables. To simplify notations, we define $\mathbf{x}(\theta) = (x_1(\theta), x_2(\theta))^T$ and $\mathbf{p}(\theta) = (p_1(\theta), p_2(\theta))^T$.

According to *Pontryagin minimum principle* [17], the optimal solution $(\mathbf{x}^*(\theta), y^*(\theta))$ to problem **P1$'$** should satisfy the following six conditions:

$$
\begin{aligned}
\dot{x}_1^*(\theta) &= \frac{\partial H\left(\mathbf{x}^*(\theta), y^*(\theta), \mathbf{p}^*(\theta), \theta\right)}{\partial p_1(\theta)} \\
&= -\beta q_{\max} y^*(\theta) \ ,
\end{aligned}
\tag{3.19}
$$

$$
\begin{aligned}
\dot{x}_2^*(\theta) &= \frac{\partial H\left(\mathbf{x}^*(\theta), y^*(\theta), \mathbf{p}^*(\theta), \theta\right)}{\partial p_2(\theta)} \\
&= d\left(q_{\max} y^*(\theta), 1 - \beta\right) f(\theta) \ ,
\end{aligned}
\tag{3.20}
$$

$$
\dot{p}_1^*(\theta) = -\frac{\partial H\left(\mathbf{x}^*(\theta), y^*(\theta), \mathbf{p}^*(\theta), \theta\right)}{\partial x_1(\theta)} = f(\theta) \ ,
\tag{3.21}
$$

$$
\dot{p}_2^*(\theta) = -\frac{\partial H\left(\mathbf{x}^*(\theta), y^*(\theta), \mathbf{p}^*(\theta), \theta\right)}{\partial x_2(\theta)} = 0 \ ,
\tag{3.22}
$$

$$H\left(\mathbf{x}^*\left(\theta\right),y^*\left(\theta\right),\mathbf{p}^*\left(\theta\right),\theta\right)\geq H\left(\mathbf{x}^*\left(\theta\right),y\left(\theta\right),\mathbf{p}^*\left(\theta\right),\theta\right),\tag{3.23}$$

$$p_1^*\left(\underline{\theta}\right)=0.\tag{3.24}$$

From (3.21) and (3.24) we can get

$$p_1^*\left(\theta\right)=F\left(\theta\right).\tag{3.25}$$

From (3.22) we know that for any $\theta\in\left[\underline{\theta},\bar{\theta}\right]$, there is

$$p_2^*\left(\theta\right)=\gamma,\tag{3.26}$$

where the γ will later be determined by using the boundary condition $x_2\left(\bar{\theta}\right)=q_{req}/Nq_{max}$.

Having determined $p_1^*\left(\theta\right)$ and $p_2^*\left(\theta\right)$, now we need to optimize the Hamiltonian with respect to $y\left(\theta\right)$. In order to derive the analytic expression of $y^*\left(\theta\right)$, we assume that data provider's type is uniformly distributed within $\left[\underline{\theta},\bar{\theta}\right]$. The probability density function is

$$f\left(\theta\right)=\frac{1}{\bar{\theta}-\underline{\theta}},\forall\theta\in\left[\underline{\theta},\bar{\theta}\right],\tag{3.27}$$

and the cumulative density function is

$$F\left(\theta\right)=\frac{\theta-\underline{\theta}}{\bar{\theta}-\underline{\theta}},\forall\theta\in\left[\underline{\theta},\bar{\theta}\right].\tag{3.28}$$

Given above assumption, the optimal production function $\rho_\beta^*\left(\cdot\right)$ can be derived via following two steps. First, we ignore the boundary constraint (3.9) and solve the unbounded $\tilde{y}_\beta\left(\cdot\right)$ that maximizes the Hamiltonian. By using the first order condition $\frac{\partial H\left(\mathbf{x}^*(\theta),y(\theta),\mathbf{p}^*(\theta),\theta\right)}{\partial y}\bigg|_{y=\tilde{y}}=0$ we get

$$\tilde{y}_\beta\left(\theta\right)=\frac{\left(\alpha_1\beta^{\alpha_2}+\alpha_3\right)\lambda^2}{4\left[\left(2\theta-\underline{\theta}\right)\beta-\gamma\right]^2 q_{max}},\forall\theta\in\left[\underline{\theta},\bar{\theta}\right].\tag{3.29}$$

It can be easily verified that

$$\frac{\partial^2 H\left(\mathbf{x}^*\left(\theta\right),y\left(\theta\right),\mathbf{p}^*\left(\theta\right),\theta\right)}{\partial y^2}\bigg|_{y=\tilde{y}}<0,\tag{3.30}$$

hence $\tilde{y}_\beta(\cdot)$ does maximize the Hamiltonian. The β-specific constant γ in (3.29) can be determined by using the monotonicity constraint (3.13) and the isoperimetric constraint (3.8). Take the derivative of $\tilde{y}_\beta(\theta)$ with respect to θ and use (3.13) we get $\gamma < \beta\underline{\theta}$. Then, plug (3.29) into the right-hand side of (3.8) and solve the equation for γ, we get

$$\gamma = \beta\bar{\theta} - \frac{1}{2}\sqrt{4(\bar{\theta}-\underline{\theta})^2\beta^2 + \frac{(\alpha_1\beta^{\alpha_2}+\alpha_3)N\lambda^2}{q_{req}}}. \tag{3.31}$$

With $\tilde{y}_\beta(\cdot)$ determined, the next step to find $\rho_\beta^*(\cdot)$ is to check whether the boundary constraint (3.9) can be satisfied.

Let us first consider a special case, that is, the data collector can offer a perfect protection of privacy, namely $\beta = 0$. In such a case, $\tilde{y}_\beta(\cdot)$ becomes a constant function, i.e.

$$y_0^*(\theta) = \frac{q_{req}}{Nq_{max}}, \quad \forall\theta \in [\underline{\theta}, \bar{\theta}]. \tag{3.32}$$

Then, according to (3.14) and (3.15), each data provider will receive zero information rent. The intuition behind this result is that when no privacy disclosure will happen, there is no privacy loss to data providers. Thus the data collector is indifferent to how each provider values his privacy, and the task of data producing is equally assigned to different providers. As for the data provider, since his type θ does not matter to the collector, namely his information advantage over the collector no longer exists, he will receive no information rent. The total payoff to the data collector is

$$U_C^*(0) = \lambda\sqrt{\alpha_3 Nq_{req}}. \tag{3.33}$$

Given $\beta = 0$, the optimal contract $\left(1-\beta^*, t_\beta^*(.), \rho_\beta^*(.)\right)$ has a very simple form, which is $\left(0, 0, \frac{q_{req}}{Nq_{max}}\right)$. However, in practice, perfect privacy protection can hardly be realized, thus such a contract is impractical. It is more important to explore the cases when privacy disclosure is inevitable.

Given $\beta \in (0,1]$, $\tilde{y}_\beta(\cdot)$ can be depicted by the curve segment shown in Fig. 3.2. As we can see, if the curve segment intersects with the line $y(\theta) = 1$ at some point, then $\tilde{y}_\beta(\cdot)$ cannot be taken as a feasible production function. Let $(\theta_c, 1)$ denote the intersection point (possibility exists), where θ_c is defined as

$$\theta_c = \frac{1}{2}(\bar{\theta}+\underline{\theta}) - \frac{1}{4\beta}\sqrt{4(\bar{\theta}-\underline{\theta})^2\beta^2 + \frac{(\alpha_1\beta^{\alpha_2}+\alpha_3)N\lambda^2}{q_{req}}}$$
$$+ \frac{\lambda}{4\beta}\sqrt{\frac{\alpha_1\beta^{\alpha_2}+\alpha_3}{q_{max}}}. \tag{3.34}$$

Fig. 3.2 Optimal production functions under different settings of q_{req}

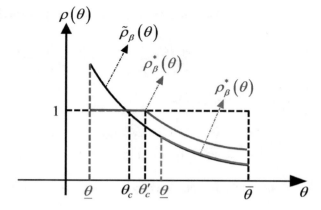

If θ_c lies outside the interval $[\underline{\theta}, \bar{\theta}]$, then $\tilde{y}_\beta(\cdot)$ is the optimal production function we are looking for. Through a simple analysis of (3.34) we learn that, as long as $q_{req} \leq Nq_{max}$, there is $\theta_c < \frac{1}{2}(\bar{\theta} + \underline{\theta}) < \bar{\theta}$. However, it is uncertain whether there is $\theta_c < \underline{\theta}$.

As defined in (3.34), given the exogenous parameters $\{\lambda, N, q_{max}, \bar{\theta}, \underline{\theta}, \alpha_1, \alpha_2, \alpha_3\}$ and β, the value of θ_c is fully determined by the collector's requirement q_{req}. With the increase of q_{req}, θ_c increases. At some point when q_{req} is higher than a threshold q_{maxreq}, θ_c will exceed $\underline{\theta}$. The threshold q_{maxreq} can determined by setting $\theta_c = \underline{\theta}$, and clearly it depends on β as follows

$$q_{maxreq}(\beta) = \frac{\lambda N q_{max}\sqrt{\alpha_1\beta^{\alpha_2} + \alpha_3}}{\lambda\sqrt{\alpha_1\beta^{\alpha_2} + \alpha_3} + 4(\bar{\theta} - \underline{\theta})\beta q_{max}} . \tag{3.35}$$

Considering that $q_{req} \in (0, Nq_{max}]$ is specified before the contract is formed, the following three situations need to be analyzed respectively.

1. If $0 < q_{req} \leq q_{maxreq}(\beta)$, then for all $\theta \in [\underline{\theta}, \bar{\theta}]$, $\tilde{y}_\beta(\theta)$ lies within the boundaries. In such a case, the optimal production function $\rho_\beta^*(\cdot)$ has exactly the formulation with $\tilde{y}_\beta(\cdot)$ as defined in (3.29). Then, by using (3.14) and (3.15) we can determine the optimal information rent function, that is

$$U_\beta^*(\theta) = \frac{(\alpha_1\beta^{\alpha_2} + \alpha_3)\lambda^2}{8\left[(2\theta - \underline{\theta})\beta - \gamma\right]} - \frac{(\alpha_1\beta^{\alpha_2} + \alpha_3)\lambda^2}{8\left[(2\bar{\theta} - \underline{\theta})\beta - \gamma\right]} . \tag{3.36}$$

2. If $q_{maxreq}(\beta) < q_{req} < Nq_{max}$, then for $\theta \in [\underline{\theta}, \theta_c]$, $\tilde{y}_\beta(\theta)$ lies outside the boundary. In such a case, we define $\rho_\beta^*(\cdot)$ as a piecewise function, i.e.

$$\rho_\beta^*(\theta) = \begin{cases} 1, & \underline{\theta} \leq \theta \leq \theta'_c \\ \dfrac{(\alpha_1\beta^{\alpha_2} + \alpha_3)\lambda^2}{4\left[(2\theta - \underline{\theta})\beta - \gamma'\right]^2 q_{max}}, & \theta'_c < \theta \leq \bar{\theta} \end{cases} \tag{3.37}$$

where γ' is determined by using (3.8) and (3.13). Specifically, γ' is given by

$$\gamma' = \beta\bar{\theta} - \frac{\lambda}{2}\sqrt{\frac{\alpha_1\beta^{\alpha_2} + \alpha_3}{q_{\max}}} + \frac{(\bar{\theta} - \underline{\theta})\,q_{req}\beta}{Nq_{max}} - \frac{\sqrt{\Delta'}}{Nq_{\max}}, \qquad (3.38)$$

where Δ' is defined as

$$\begin{aligned} \Delta' = {} & (\bar{\theta} - \underline{\theta})^2 \left(Nq_{\max} - q_{req}\right)^2 \beta^2 \\ & + N\lambda\beta\sqrt{(\alpha_1\beta^{\alpha_2} + \alpha_3)\,q_{\max}}\,(\bar{\theta} - \underline{\theta})\left(Nq_{\max} - q_{req}\right) \end{aligned}. \qquad (3.39)$$

Based on the formulation of γ', θ'_c can be determined by

$$\theta'_c = \frac{1}{2}\underline{\theta} + \frac{\gamma'}{2\beta} + \frac{\lambda}{4\beta}\sqrt{\frac{\alpha_1\beta^{\alpha_2} + \alpha_3}{q_{\max}}}. \qquad (3.40)$$

A geometric interpretation of (3.37) is given below. As shown in Fig. 3.2, the area under the black curve segment is proportional to the collector's requirement q_{req}. When θ_c lies on the right side of $\underline{\theta}$, the curve segment can be divided into two subsegments, namely the one lies on the left side of the point $(\theta_c, 1)$ and the one lies on the right side. For points on the left-hand segment, we has to "pull" them down until they reach the boundary. By doing so, the area between the original segment and the boundary is discarded. In order to keep the total area unchanged, points on the right-hand segment must be "pushed up", and those who lie close to $(\theta_c, 1)$ may be pushed up to the boundary. For a given β, as q_{req} decreases, the whole curve segment moves towards the left, which means fewer points need to be pulled down. By the time q_{req} decreases to $q_{maxreq}(\beta)$, the curve intersects with the boundary at $(\underline{\theta}, 1)$. In such a case, no point needs to be pulled down, and this is when the piecewise $\rho_\beta^*(\cdot)$ degenerates to that defined in (3.29). On the contrary, as q_{req} increases, the whole curve segment moves towards the right, and more points lie above the boundary. Consequently, the pulling-down operation causes a larger loss in area, which means points on the right-hand segment should be pushed higher. In an extreme case, all the points on the left-hand segment are pushed to the boundary. This is exactly the third case that we will discuss later.

With $\rho_\beta^*(\cdot)$ defined in (3.37), we can derive the optimal information rent function $U_\beta^*(\cdot)$ by using (3.14) and (3.15), that is

$$U_\beta^*(\theta) = \begin{cases} -\beta q_{\max}\theta + \Gamma_\beta, & \theta \in [\underline{\theta}, \theta'_c] \\ \dfrac{(\alpha_1\beta^{\alpha_2} + \alpha_3)(\bar{\theta} - \theta)\lambda^2\beta}{4[(2\theta - \underline{\theta})\beta - 2\gamma'][(2\bar{\theta} - \underline{\theta})\beta - 2\gamma']}, & \theta \in (\theta'_c, \bar{\theta}] \end{cases} \qquad (3.41)$$

where Γ_β is defined as

$$\Gamma_\beta = \beta q_{\max} \theta'_c$$
$$+ \frac{\left(\alpha_1 \beta^{\alpha_2} + \alpha_3\right) \left(\bar{\theta} - \theta'_c\right) \lambda^2 \beta}{4\left[\left(2\theta'_c - \underline{\theta}\right)\beta - 2\gamma'\right]\left[\left(2\bar{\theta} - \underline{\theta}\right)\beta - 2\gamma'\right]}. \tag{3.42}$$

3. If $q_{req} = N q_{\max}$, similar to above case, θ_c lies to the right side of $\underline{\theta}$, hence the optimal production function $\rho^*_\beta(\cdot)$ has the same form as that defined in (3.37). But in this case, there is $\theta'_c = \bar{\theta}$, and (3.37) becomes a constant function, i.e.

$$\rho^*_\beta(\theta) = 1, \ \forall \theta \in \left[\underline{\theta}, \bar{\theta}\right]. \tag{3.43}$$

Again, by using (3.14) and (3.15) we get the optimal information rent function, which is defined as

$$U^*_\beta(\theta) = \left(\bar{\theta} - \theta\right)\beta q_{\max}. \tag{3.44}$$

Then the optimal transfer function $t^*_\beta(\cdot)$ can be written as

$$t^*_\beta(\theta) = U^*_\beta(\theta) + \theta \beta q_{\max} \rho^*_\beta(\theta) = \beta q_{\max} \bar{\theta}. \tag{3.45}$$

This result coincides with the intuition that when different data providers provide the same amount of data, they will be paid equally.

Above we have discussed how to design the optimal production function $\rho^*_\beta(\cdot)$ and optimal information rent function $U^*_\beta(\cdot)$ for a given privacy protection level. As we have clarified, different forms of these two functions should be adopted in accordance with different values of q_{req}. It should be noted that as q_{req} approaches $q_{\max req}(\beta)$ (or $N q_{\max}$), the piecewise production function defined in (3.37) will degenerate to a smooth form.

3.3.4 Determining the Optimal Privacy Protection Level

The production function $\rho^*_\beta(\cdot)$ and the information rent function $U^*_\beta(\cdot)$ derived in the above subsection are optimal for a given privacy protection level. In other words, both the functions are parameterized by β. With these optimal functions, the data collector can determine the optimal privacy protection level by solving the ordinary optimization problem **P2**. Similar to previous discussions, in this subsection we study the optimization problem by considering two cases, namely $0 < q_{req} < N q_{\max}$ and $q_{req} = N q_{\max}$.

1. $0 < q_{req} < Nq_{max}$

As discussed in Sect. 3.3.3, for each $\beta \in (0, 1]$ and $q_{req} \in (0, Nq_{max}]$, there exists a threshold $q_{maxreq}(\beta)$ which determines the maximal data requirement that can be realized by the production function defined in (3.29). According to (3.35), $q_{maxreq}(\beta)$ monotonically decreases with β. Thus, given $q_{req} \in (0, Nq_{max}]$, there exists a threshold $q_{maxreq}^{-1}(q_{req})$, where $q_{maxreq}^{-1}(\cdot)$ denotes the inverse function of $q_{maxreq}(\cdot)$, such that when $\beta \leq q_{maxreq}^{-1}(q_{req})$, $\rho_\beta^*(\theta)$ takes the form defined in (3.29), and when $q_{maxreq}^{-1}(q_{req}) < \beta \leq 1$, $\rho_\beta^*(\theta)$ takes the form defined in (3.37). Notice that when $q_{req} = Nq_{max}$, there is $q_{maxreq}^{-1}(q_{req}) = 0$. We will discuss this special case later.

Given $q_{req} \in (0, Nq_{max})$, the data collector's expected payoff $U_C(\beta)$ is defined as follows:

(i) If $\beta = 0$, as we've discussed in Sect. 3.3.3, there is $U_C(\beta) = \lambda\sqrt{\alpha_3 Nq_{req}}$.

(ii) If $0 < \beta \leq q_{maxreq}^{-1}(q_{req})$, substituting (3.29) and (3.36) into the objective function of problem **P2** and calculating the integral yields

$$U_C(\beta) = \frac{(\alpha_1\beta^{\alpha_2} + \alpha_3) N\lambda^2}{8(\bar\theta - \underline\theta)\beta} \ln \frac{(\bar\theta - \underline\theta)\beta + \frac{1}{2}\Delta}{(\underline\theta - \bar\theta)\beta + \frac{1}{2}\Delta}$$
$$+ \left(\frac{1}{2}\Delta - \beta\bar\theta\right) q_{req},$$

(3.46)

where $\Delta = \sqrt{4(\bar\theta - \underline\theta)^2\beta^2 + \frac{(\alpha_1\beta^{\alpha_2} + \alpha_3)N\lambda^2}{q_{req}}}$.

(iii) If $q_{maxreq}^{-1}(q_{req}) < \beta \leq 1$, substituting (3.37) and (3.41) into the objective function of problem **P2** and calculating the integral yields

$$U_C(\beta) = \frac{(\alpha_1\beta^{\alpha_2} + \alpha_3) N\lambda^2}{8(\bar\theta - \underline\theta)\beta} \ln \frac{(2\bar\theta - \underline\theta)\beta - \gamma'}{(2\theta'_c - \underline\theta)\beta - \gamma'}$$
$$+ N\left(\lambda\sqrt{(\alpha_1\beta^{\alpha_2} + \alpha_3) q_{max}} - \beta q_{max}\theta'_c\right)\frac{\theta'_c - \underline\theta}{\bar\theta - \underline\theta}$$

(3.47)

$$+ \frac{(\alpha_1\beta^{\alpha_2} + \alpha_3)(\theta'_c - \bar\theta) N\lambda^2\gamma'}{4(\bar\theta - \underline\theta)[(2\bar\theta - \underline\theta)\beta - \gamma'][(2\theta'_c - \underline\theta)\beta - \gamma']},$$

where γ' and θ'_c are defined in (3.38) and (3.40) respectively.

It can be verified that as β approaches 0, (3.46) approaches (3.33). And as β approaches q_{maxreq}^{-1} from right-hand side, (3.47) approaches (3.46). Thus, though described in a piecewise form, the collector's payoff changes continuously with β. Let β^* denote the probability of privacy disclosure that maximizes the collector's payoff. Since both β and $U_C(\beta)$ are bounded, the existence of β^* is guaranteed. From a practical perspective, neither $\beta^* = 0$ nor $\beta^* = 1$ is desirable. If β^* does can be found in the interior, the following two conditions must hold:

$$\left. \frac{dU_C\,(\beta)}{d\beta} \right|_{\beta=\beta^*} = 0 \,, \tag{3.48}$$

$$\left. \frac{d^2 U_C\,(\beta)}{d\beta^2} \right|_{\beta=\beta^*} < 0 \,. \tag{3.49}$$

Due to the complicated form of $U_C\,(\beta)$, it is hardly to derive the analytic form of β^* from (3.48). Instead, we propose a simple yet useful method to approximately determine the optimal protection level. Suppose that the data collector employs some k-anonymity algorithm [3] to protect data providers' privacy. For a given k, the probability of privacy disclosure can be roughly defined as $\beta = \frac{1}{k}$. Since the total number of collected data records is limited, k can only be chosen from a finite set, e.g. $\{2, \cdots, Nq_{\max}\}$. Given $q_{req} \in (0, Nq_{\max})$ ($q_{req} < Nq_{\max}$), the optimal k can be determined in a following way. For each possible k, the collector first checks whether the condition $q_{req} \le q_{\text{maxreq}}\left(\frac{1}{k}\right)$ holds. If it does, the collector computes his expected payoff $U_C\left(\frac{1}{k}\right)$ by using (3.46). Otherwise, the payoff is computed according to (3.47). After obtaining all the possible payoffs, the collector can decide which k is optimal.

2. $q_{req} = Nq_{\max}$

As discussed in Sect. 3.3.3, when the collector requires the maximal data utility, i.e. $q_{req} = Nq_{\max}$, different data providers provide the same amount of data and receive the same transfer. In such a case, the collector's payoff is

$$U_C\,(\beta) = N\left[\lambda\sqrt{(\alpha_1\beta^{\alpha_2} + \alpha_3)\,q_{\max}} - \beta\bar{\theta}q_{\max}\right] \,. \tag{3.50}$$

Note that all the parameters, except λ, in the right-hand side of above equation are generally fixed. Thus, whether there exists a $\beta^* \in (0, 1)$ fully depends on λ. Later, by conducting numerical simulations, we will discuss how λ influences the choice of β^*.

3.3.5 Non-optimal Contracts

The contract proposed above is the optimal solution to problem **P**, i.e., among all the feasible contracts, it should bring the collector the maximal payoff. Despite the fact that it is impossible to explicitly compare this contract to all the other feasible contracts, here we propose a simple-formed contract, which we refer to as *linear-production contract*, with the purpose of obtaining more insight of the optimal contract. The linear-production contract is designed as follows.

Given $\beta \in (0, 1]$, the production function $\hat{\rho}_\beta\,(\cdot)$ is defined as

$$\hat{\rho}_\beta\,(\theta) = \left(\theta - \underline{\theta}\right)\kappa + 1 \,, \tag{3.51}$$

where κ is defined as

$$\kappa = \frac{2\left(q_{req} - Nq_{\max}\right)}{\left(\bar{\theta} - \underline{\theta}\right) Nq_{\max}} . \tag{3.52}$$

This linear production function implies that a data provider who does not care about privacy (i.e. $\theta = \underline{\theta}$) should hand over all his data, and for a data provider who cares about privacy, the data utility he contributes should be proportional to his privacy preference. The information rent function is defined as

$$\hat{U}_\beta\left(\theta\right) = -\frac{1}{2}\beta q_{\max}\kappa\theta^2 - \left(1 - \kappa\underline{\theta}\right)\beta q_{\max}\theta$$

$$+ \left(\frac{1}{2}\kappa\bar{\theta}^2 - \kappa\underline{\theta}\bar{\theta} + \bar{\theta}\right)\beta q_{\max} . \tag{3.53}$$

It can be verified that if the collector has a relatively high requirement on data, i.e.

$$\frac{1}{2}Nq_{\max} \le q_{req} \le Nq_{\max} , \tag{3.54}$$

then $\left(\hat{U}_\beta\left(\cdot\right), \hat{\rho}_\beta\left(\cdot\right)\right)$ is a feasible solution to problem **P1′**.

Substitute (3.51) and (3.53) into the objective function of problem **P1**, then we get

$$\hat{U}_C\left(\beta\right) = \frac{2N\lambda\sqrt{\left(\alpha_1\beta^{\alpha_2} + \alpha_3\right) q_{\max}}}{3\left(\bar{\theta} - \underline{\theta}\right)\kappa}\left\{\left[\left(\bar{\theta} - \underline{\theta}\right)\kappa + 1\right]^{\frac{3}{2}} - 1\right\}$$

$$- \frac{\left(\bar{\theta}^3 - \underline{\theta}^3\right)\kappa Nq_{\max}\beta}{6\left(\bar{\theta} - \underline{\theta}\right)} \tag{3.55}$$

$$- \left(\frac{1}{2}\kappa\bar{\theta}^2 - \kappa\bar{\theta}\underline{\theta} + \bar{\theta}\right) N\beta q_{\max}$$

Similar to the case of optimal contract, it is quite difficult to derive the analytic form of β^* that maximizes (3.55). Considering that this contract is proposed for comparison purpose, we use a experimental method to approximately determine the value of β^* for both the optimal contract and the linear-production contract. More details will be presented in Sect. 3.4.1.1.

3.4 Contract Analysis and Simulation

In the previous section we have presented an elaborate description of the design of optimal contract. Analytic forms of the production function $\rho_\beta^*(\cdot)$ and information rent function $U_\beta^*(\cdot)$, which are optimal for a given β, are proposed. The expected payoff to the data collector is explicitly formulated as a function of β. Though we do not provide an explicit formulation of the optimal privacy protection level, we can utilize the derived formulation of $U_C(\beta)$ to provide a general insight into the trade-off between privacy protection and utility preserving.

In this section, by conducting numerical simulations, we qualitatively analyze how the optimal privacy protection level relates to the collector's requirement on data utility and the exogenously determined value of data. Moreover, in order to evaluate whether the optimal contract can bring the collector a good payoff, we conduct real data experiments and make a comparison of the two types of contracts proposed in Sect. 3.3. In the following part, we first describe how we determine the optimal privacy protection level via simulations. Then based on simulation results, we present a qualitative analysis of the optimal contract. After that, we introduce the settings of real data experiments and present the results.

3.4.1 Contract Analysis

3.4.1.1 Determining the Optimal Privacy Protection Level Experimentally

To observe how the two parameters q_{req} and λ influence the choice of privacy protection level in the optimal contract, we conduct the following simulations. First, we set those invariable parameters as follows: $N = 3000$, $q_{max} = 10$, $\alpha_1 = 0.4804$, $\alpha_2 = 0.2789$, $\alpha_3 = 1 - \alpha_1$, $\bar{\theta} = 1$, and $\underline{\theta} = 0$. Then, for each pair of $q_{req} \in \left\{ \frac{1}{20} N q_{max}, \frac{2}{20} N q_{max}, \cdots, N q_{max} \right\}$ and $\lambda \in \{0.1, 0.2, \cdots 0.9, 1, 2, , \cdots, 100\}$, we compute a group of $\{U_C(\beta)\}$, each of which corresponds to a $\beta \in \left\{ \frac{0}{100}, \frac{1}{100}, \cdots, \frac{100}{100} \right\}$. For each β, we first compare q_{req} with $q_{maxreq}(\beta)$. Then based on the comparison result, $U_C(\beta)$ is computed according to (3.33), (3.46) or (3.47). After that, the maximal $U_C^*(\beta)$ is picked from $\{U_C(\beta)\}$, and the corresponding privacy protection level $\delta^* \triangleq 1 - \beta^*$ is recorded. As for the linear-production contract proposed in Sect. 3.3.5, the optimal privacy protection level is determined in a similar way.

3.4.1.2 Data Requirement and Privacy Protection

As discussed in Sect. 3.3.3, how the optimal contract should be formed largely depends on the collector's requirement on data. From the results shown in Fig. 3.3

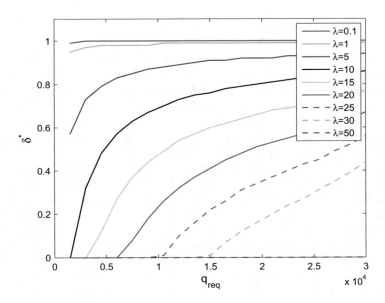

Fig. 3.3 Relationship between the optimal privacy protection level and the requirement on total data utililty

we can see that, for a given λ, δ^* increases with q_{req}, as long as λ is neither too high nor too low. This phenomenon implies that if the data collector wants to get more data from data providers, he should offer better protection for data providers' privacy. Or in other words, knowing that his privacy can be better protected, the data provider will feel less unsafe to hand over his private data, thus he is willing to provide more data.

To better understand above implication, we rewrite the collector's expected payoff $U_C(\beta)$ as a sum of two terms, i.e.

$$U_C(\beta) = S(\beta) - T(\beta) \,, \tag{3.56}$$

where $S(\beta)$ denotes the expected income, i.e.

$$S(\beta) = N \int_{\underline{\theta}}^{\bar{\theta}} \lambda \sqrt{(\alpha_1 \beta^{\alpha_2} + \alpha_3) q_{max} \rho_\beta^*(\theta)} f(\theta) \, d\theta, \tag{3.57}$$

and $T(\beta)$ denotes the expected transfer, i.e.

$$T(\beta) = N \int_{\underline{\theta}}^{\bar{\theta}} \left[U_\beta^*(\theta) + \beta \theta q_{max} \rho_\beta^*(\theta) \right] f(\theta) \, d\theta. \tag{3.58}$$

During the simulations, we compute $S(\beta)$ and $T(\beta)$ together with $U_C(\beta)$. As shown in Fig. 3.4, with λ being fixed at a moderate value (e.g. $\lambda = 15$), for any

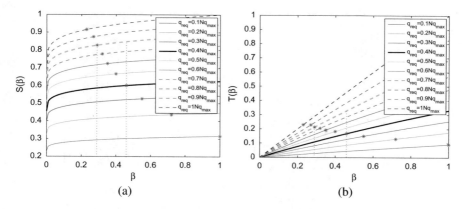

Fig. 3.4 An illustration of how data collector's income and transfer change with the privacy protection level: (**a**) expected income $S(\beta)$; (**b**) expected transfer $T(\beta)$. The plots, which denote the optima, are obtained via the simulations described in Sect. 3.4.1.1 where λ is set to 15. Values of $S(\beta)$ shown in the figure have been normalized by dividing original value by the maximum among all results. Values of $T(\beta)$ have been normalized in a similar way. Red stars denote the income (or transfer) at the optimum, i.e. $(\beta^*, S(\beta^*))$ (or $(\beta^*, T(\beta^*))$)

given q_{req}, both the income and the transfer increase with β. This coincides with the intuition that when privacy protection level decreases, the collector can obtain more benefit from the less anonymized data, meanwhile, data providers face a higher risk of privacy disclosure, hence they require more transfer to compensate the privacy loss. From Fig. 3.4 we can see that, compared to the income $S(\beta)$, the transfer $T(\beta)$ is more sensitive to β. And as q_{req} becomes higher, $T(\beta)$ grows faster with β, while $S(\beta)$ grows at almost the same rate. According to (3.2), the marginal value of data decrease with the utility of anonymitized data which, according to (3.1), grows slower as the utility of collected data increases. This may explain why $S(\beta)$ is insensitive to β. While as for $T(\beta)$, it is roughly proportional to β, which means even a small change of β can be captured by $T(\beta)$.

Suppose that currently the collector's data requirement is $q_{req} = 0.4Nq_{max}$, and the optimal privacy protection level he adopts is about 0.57. When the collector has a higher requirement, say $q_{req} = 0.8Nq_{max}$, he has to pay much more transfer if he sticks with original privacy protection level. However, if the collector chooses a higher protection level, despite that he'll lose a small amount of income, the transfer he needs to pay can be largely reduced. Therefore, when the collector desires data of high utility, he should put more effort to protect data providers' privacy.

Figure 3.3 also shows that when q_{req} is relatively small and λ is large, the collector does not need to take care of data providers' privacy. This is because that when data is very valuable, the income from the data is far beyond sufficient to compensate data providers' privacy loss, thus there is no need to take privacy protection measures. However, as the collector has collected more data to meet a higher q_{req}, the marginal value of data decreases, and the income may be insufficient to compensate the privacy loss. Therefore, the collector should take

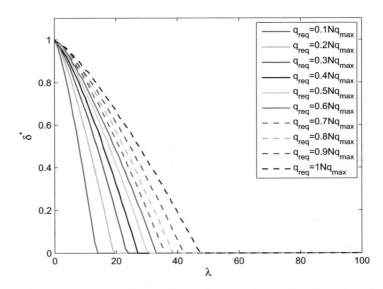

Fig. 3.5 Relationship between the optimal privacy protection level and the value of data

privacy protection measures, so that the transfer paid to data providers can be reduced to an affordable level.

3.4.1.3   The Value of Data and Privacy Protection

The parameter λ in data collector's income function (3.2) indicates whether the data is valuable to the collector. From the simulation results shown in Fig. 3.5 we can see that, when λ is quite small ($\lambda < 1$), the optimal privacy protection level equals 1, which means the data collector must offer a perfect protection of privacy. The reason why this result appears is that we have defined the privacy parameter θ takes values from 0 to 1. Considering that θ can be interpreted as the unit cost that a data provider spends on producing the data, the transfer that the data collector pays to the provider should be at an equivalent level. Then when $\theta \in [0, 1]$ and $\lambda < 1$, the benefit that the collector gets from the data may be even less than the provider's cost, which means the collector cannot afford any compensation for data providers' privacy loss. As a result, providing perfect privacy protection may be the only feasible choice for the collector.

As shown in Fig. 3.5, the optimal privacy protection level decreases as λ increases. This implies that as data becomes more valuable, sacrificing data utility for privacy protection becomes less beneficial to the collector. In such cases, though increasing privacy protection level can reduce the transfer paid to data providers, the resulting decrease of data utility will cause a larger loss to the collector. When λ is quite large, data is so valuable to the collector that even a minor decrease in data utility, which is caused by a weak anonymization, will cause a large loss to the collector. As a result, the collector prefers to do nothing to protect privacy. From Fig. 3.5 we can see that, as q_{req} becomes higher, in a larger range of λ, protecting privacy is more preferred by the collector than providing no protection. This result coincides with the observation we've got in Sect. 3.4.1.2, that is, a better protection of privacy is required if the collector wants to get data of higher utility.

More insight about how the parameter λ influences the design of privacy protection level can be obtained by analyzing the second case discussed in Sect. 3.3.4, i.e. $q_{req} = Nq_{max}$. As mentioned earlier, whether the data collector's payoff can reach its maximum at an interior β can be determined by evaluating the derivation as follows.

When λ meets the following condition

$$\lambda \geq \max_{\beta \in [0,1]} \frac{2\bar{\theta}\beta^{1-\alpha_2}}{\alpha_1\alpha_2}\sqrt{(\alpha_1\beta^{\alpha_2}+\alpha_3)\,q_{max}} = \frac{2\bar{\theta}\sqrt{q_{max}}}{\alpha_1\alpha_2}, \tag{3.59}$$

there is $\frac{dU_C(\beta)}{d\beta} \leq 0$, and $\frac{dU_C(\beta)}{d\beta} = 0$ iff $\beta = 1$. In such a case, the collector's payoff increases as the privacy protection level decreases, thus $\beta^* = 1$.

When $0 < \lambda < \frac{2\bar{\theta}\sqrt{q_{max}}}{\alpha_1\alpha_2}$, $U_C(\beta)$ reaches its maximum at an interior $\beta^* \in (0,1)$ which satisfies $\frac{dU_C(\beta)}{d\beta}\Big|_{\beta=\beta^*} = 0$. It can be verified that the second order condition $\frac{dU_C^2(\beta)}{d\beta^2}\Big|_{\beta=\beta^*} < 0$ also holds. From Fig. 3.4 we know that, as the privacy protection level increases (i.e. β decreases), both the income and the transfer decreases. When λ is relatively small, the reduced transfer caused by one-unit increase of protection level is comparable with the corresponding income loss. At some point, a small increase of protection level causes no change to the payoff, that's when the payoff is maximized. Moreover, notice that $U_C(1) = N\sqrt{q_{max}}\left(\lambda - \bar{\theta}\sqrt{q_{max}}\right)$ and $\bar{\theta}\sqrt{q_{max}} < \frac{2\bar{\theta}\sqrt{q_{max}}}{\alpha_1\alpha_2}$. Hence, if $\lambda < \bar{\theta}\sqrt{q_{max}}$, the data collector cannot get a positive payoff unless a certain level of privacy protection can be realized. From Fig. 3.5 we can see that, as λ becomes smaller, β^* moves towards 0. This phenomena implies that as the data becomes less valuable to the collector, the collector has to put more effort to protect data providers' privacy, so that a low transfer will be accepted by data providers and the collector can still keep his payoff stay at a certain level.

3.4.2 Experiments on Real-World Data

3.4.2.1 Dataset and Anonymization Configurations

To evaluate the performance of the contracts in a context where anonymization is performed on real data, we conduct experiments on the Adult data set [18], which is widely used in the study of data anonymization. The original data set consists of

32,561 records from a census database, and each record consists of 15 attributes. After removing records with unknown values, we randomly choose 30,000 records for experiment. Similar to previous study on anonymization [4, 19], only nine attributes, namely *age*, *workclass*, *education*, *marital-status*, *occupation*, *race*, *sex*, *native-country*, and *salary-class*, are kept for experiment.

To perform anonymization, we develop a java project based on the open source anonymization framework ARX [20], which supports different types of privacy criteria and provides multiple methods for measuring information loss [1]. Here we choose the most widely applied privacy criterion, i.e., k-anonymity, to conduct experiments. A simple explanation to this privacy criterion is that after anonymization, the probability that an individual being re-identified by an attacker is no higher than $1/k$. Hence, if the k-anonymity criterion is met by the anonymized data, the realized privacy protection level can be defined as $\delta \triangleq 1 - \frac{1}{k}$.

After anonymization, the utility of data decreases. The decrease of utility, also referred to as *information loss*, can be measured in different ways. Here we choose the *precision* metric [3], which ranges from 0 to 1. Intuitively, if a larger k is chosen as the privacy criterion, the information loss will becomes higher. In order to quantitatively analyze how the information loss changes with k, we conduct a group anonymization experiments on aforementioned data set. All the nine attributes are treated as quasi-identifiers, namely each of them can be generalized according to a domain generalization hierarchy [19]. For each $k \in \{2, \cdots, 50\}$, we run the anonymization program and record the reported information loss IL. Experiment results are shown in Fig. 3.6. By using the curve fitting toolbox provided in MATLAB, we formulate IL as a power function of k. Then, by defining $IL = \frac{q-d(q,\delta)}{q}$, which means IL is interpreted as the ratio of the decreased utility to that of the original data, we get the formulation defined in (3.1).

3.4.2.2   Contract Simulation

To demonstrate the superiority of the optimal contract over the linear-production contract, we conduct multiple experiments to simulate data providers' response to different contracts, and check whether the optimal contract can bring the data collector a higher payoff. Experiments are configured in the following way. First, we randomly divide the 30,000 records into N groups, where N is set to 3000, 300, and 30 respectively. Each group of records corresponds to a data provider. That is to say, we set $q_{max} = 10, 100, 1000$ respectively. The privacy parameter θ of each data provider is set by uniformly sampling in the interval $[0, 1]$. The rest parameters are set as $\lambda = 15$, $\alpha_1 = 0.4804$, $\alpha_2 = 0.2789$, $\alpha_3 = 0.5196$, and $q_{req} = \frac{m}{20}Nq_{max}$ ($m = 10, 11, \cdots, 20$).

Given the value of N and the value of q_{req}, the maximal payoff that the data collector can get from a certain contract is computed as follows. First, we determine the optimal privacy protection level β^* by using the method described in Sect. 3.4.1.1. Then, based on the production function $\rho_{\beta^*}^*(\cdot)$ defined in the contract, we determine the number of records that each data provider i will provide.

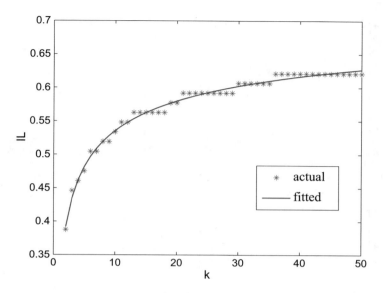

Fig. 3.6 Relationship between the privacy criterion and information loss. The information loss IL is measure by *precision* [3]. The blue curve is fitted by using the data denoted by red stars. By using MATLAB curve fitting toolbox, we choose a power function to formulate the fitted curve, which is $IL = -0.4804k^{-0.2789} + 0.7883$. From the reported R-square (coefficient of determination) index, which is 0.9896, we know that such formulation is appropriate

Let n_i denote the number of records and θ_i denote the provider's type. We set $n_i = \left\lceil \rho_{\beta*}^* (\theta_i) q_{\max} \right\rceil$, where $\lceil a \rceil$ denote the smallest integer that is no less than a. Based on n_i and the information rent function $U_{\beta*}^* (\cdot)$, the information rent u_i paid to provider i can be determined. After above computation, we construct a new data set by randomly choosing n_i records from the ten records corresponding to each provider i. To run anonymization experiments on this data set, we set $k = \left\lceil \frac{1}{\beta*} \right\rceil$. Then, based on the reported information loss and each provider's (n_i, u_i), we can determine the collector's payoff U_C^*. Considering that records in the new data are randomly chosen, for a given q_{req} and a contract, we repeat above procedure five times and report the average results.

3.4.2.3 Comparison Results

Simulation results are shown in Fig. 3.7. As we can see, in all settings of q_{req}, the optimal contract exhibits a better performance than the linear-production contract. As shown in Fig. 3.7a, c and e, in all settings of q_{req}, the optimal contract can bring the collector a higher payoff than, if not equal to, that brought by the linear-production contract, especially when q_{req} is close to $0.5Nq_{max}$. As q_{req} increases,

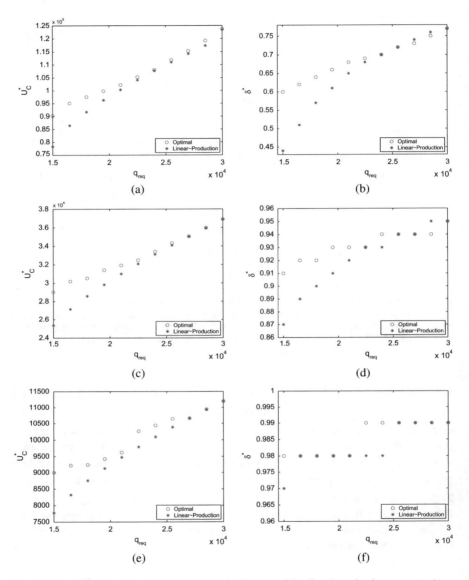

Fig. 3.7 Performance evaluation of the optimal contract and the linear-production contract: (**a, c, e**) data collector's payoff; (**b, d, f**) optimal privacy protection level, $\delta^* = 1 - \beta^*$. (**a**) $N = 3000$. (**b**) $N = 3000$. (**c**) $N = 300$. (**d**) $N = 300$. (**e**) $N = 30$. (**f**) $N = 30$

the difference between the two contracts, in terms of payoff, becomes insignificant. On the other hand, Fig. 3.7b, d and f show that when $q_{req} < 0.75Nq_{max}$, the optimal contract can offer the data providers a better protection of privacy. However, as q_{req} becomes quite high, the optimal contract can only realize a similar, even lower, privacy protection level as that realized by linear-production contract. It should be noted that a lower privacy protection level does not mean the optimal contract is worse than the linear-production contract, since the optimal contract is designed to maximize the data collector's expected payoff rather than maximizing the privacy protection level. Also, from Fig. 3.7f we can see that when the number of data providers is quite small ($N = 30$), the optimal privacy protection level approaches to 1 in all settings of q_{req}. This result can be explained in a following way. In the setting where $N = 30$, each data provider owns 1000 records. In order to meet the collector's requirement, say $q_{req} = 20,000$, on average each data provider has to provide more than 600 records. The data providers will show great concern about privacy when they are asked to provide so many private data. Thus it is necessary for the collector to offer strong protection to the providers' privacy.

To understand why the optimal contract loses its advantage when q_{req} is high, we can recall the geometrical interpretation presented in Sect. 3.3.3. As illustrated in Fig. 3.2, a high q_{req} means a large part of the curve segment defined by the production function lies on the boundary. And as q_{req} becomes higher, the rest part of the curve becomes more "flat". As for the linear-production contract, it uses a liner production function. According to (3.52), when q_{req} approaches Nq_{max}, the production decreases with θ at a very low rate. To sum up, when q_{req} is close to the maximum Nq_{max}, the two types of contracts will make no obvious difference in their production functions, hence they exhibits similar performance. In addition to above results, it should be noted that the linear-production contract can be applied only when $0.5Nq_{max} \leq q_{req} \leq Nq_{max}$, while the optimal contract can also be applied to cases when q_{req} is low. In that sense, the optimal contract is more practical than the linear-production contract.

3.5 Conclusion

To deal with the information asymmetry problem emerging in private data collecting, in this chapter we proposed a contract theoretic approach to help the data collector make a rational decision on how to pay the data providers. Considering that the data collector also needs to carefully adjust the privacy protection level, we treated the privacy protection level as a contract item, and explicitly solved the optimal production functions and information rent functions for any given protection level. We've shown that as the collector's requirement on data changes, the optimal functions may be formed in a different way. As for the optimal privacy protection level, we've analyzed how it should be adjusted when the collector faces a different requirement on data utility or has a new valuation of data. Such analysis can provide a practical guidance for private data collecting.

The optimal contract proposed in this chapter is mainly based on the assumptions we've made on data collector's income function as well as the relationship between data utility and privacy protection level. Whether there are more reasonable formulations of these two functions needs to be further investigated. Besides, in our study we have assumed that the distribution of data providers' privacy preference is known to the collector. In future work, we will study the contract design problem in a context where the distribution knowledge is unavailable to the collector. Moreover, currently we assume that the data provider's privacy parameter is pre-specified by the *nature*, yet it is important to explore practical ways to quantify individual's preference on privacy. Whether we can learn one's valuation of his privacy from one's historical behavior is a problem worth further studying.

Appendix

According to (3.6), for any $\left(\theta, \tilde{\theta}\right) \in \left[\underline{\theta}, \bar{\theta}\right]^2$, the following two inequalities hold:

$$t\left(\theta\right) - \beta\theta\rho\left(\theta\right)q_{\max} \geq t\left(\tilde{\theta}\right) - \beta\theta\rho\left(\tilde{\theta}\right)q_{\max}, \tag{3.60}$$

$$t\left(\tilde{\theta}\right) - \beta\tilde{\theta}\rho\left(\tilde{\theta}\right)q_{\max} \geq t\left(\theta\right) - \beta\tilde{\theta}\rho\left(\theta\right)q_{\max}. \tag{3.61}$$

Adding (3.60) and (3.61) yields

$$\left(\tilde{\theta} - \theta\right)\left(\rho\left(\theta\right) - \rho\left(\tilde{\theta}\right)\right)\beta q_{\max} \geq 0. \tag{3.62}$$

Above inequality should hold for any $\beta \in [0, 1]$, which means $\rho\left(\cdot\right)$ has to be a non-increasing function of θ. Furthermore, (3.62) implies that both $\rho\left(\cdot\right)$ and $t\left(\cdot\right)$ are differentiable almost everywhere. Hence, we can restrict the analysis to piecewise differentiable functions. Given θ, (3.60) implies that the function $g\left(\tilde{\theta}\right) \triangleq t\left(\tilde{\theta}\right) - \theta\beta q_{\max}\rho\left(\tilde{\theta}\right)$ reaches its maximum at $\tilde{\theta} = \theta$, thus θ must satisfy the following two conditions:

$$\frac{dt\left(\theta\right)}{d\theta} - \beta q_{\max}\theta\frac{d\rho\left(\theta\right)}{d\theta} = 0, \tag{3.63}$$

$$\frac{d^2t\left(\theta\right)}{d\theta^2} - \beta q_{\max}\theta\frac{d^2\rho\left(\theta\right)}{d\theta^2} \leq 0. \tag{3.64}$$

By differentiating (3.63), (3.64) can be written as:

$$-\frac{d\rho\left(\theta\right)}{d\theta} \geq 0. \tag{3.65}$$

The (3.63) and (3.65) constitute the local incentive constraints. Then, by using (3.63) we can write the data provider's information rent as

$$t\left(\theta\right) - \theta\beta q_{\max}\rho\left(\theta\right) = t\left(\tilde{\theta}\right) - \theta\beta q_{\max}\rho\left(\tilde{\theta}\right) +$$

$$\beta q_{\max}\int_{\tilde{\theta}}^{\theta}\left[\rho\left(\tilde{\theta}\right) - \rho\left(\tau\right)\right]d\tau\ . \tag{3.66}$$

The non-increasing property (3.65) ensures that the third item in the right-hand side of above equation is non-negative, which means the local incentive constraints imply also the global incentive constraints. Hence, we can reduce the infinity of incentive constraints in (3.6) to a differential equation (3.63) and a monotonicity constraint (3.65).

References

1. B. Fung, K. Wang, R. Chen, and P. S. Yu, "Privacy-preserving data publishing: A survey of recent developments," *ACM Comput. Surv.*, vol. 42, no. 4, pp. 1–53, 2010.
2. R. Agrawal and R. Srikant, "Privacy-preserving data mining," *SIGMOD Rec.*, vol. 29, no. 2, pp. 439–450, 2000.
3. L. SWEENEY, "Achieving k-anonymity privacy protection using generalization and suppression," *International Journal of Uncertainty, Fuzziness and Knowledge-Based Systems*, vol. 10, no. 05, pp. 571–588, 2002.
4. A. Machanavajjhala, J. Gehrke, D. Kifer, and M. Venkitasubramaniam, "L-diversity: privacy beyond k-anonymity," in *Data Engineering, 2006. ICDE '06. Proceedings of the 22nd International Conference on*, April 2006, pp. 24–24.
5. N. Li, T. Li, and S. Venkatasubramanian, "t-closeness: Privacy beyond k-anonymity and l-diversity." in *ICDE*, vol. 7, 2007, pp. 106–115.
6. C. C. Aggarwal and S. Y. Philip, *A general survey of privacy-preserving data mining models and algorithms*. Springer, 2008.
7. S. Matwin, "Privacy-preserving data mining techniques: Survey and challenges," in *Discrimination and Privacy in the Information Society*. Springer, 2013, pp. 209–221.
8. A. Acquisti, C. R. Taylor, and L. Wagman, "The economics of privacy," *Journal of Economic Literature*, vol. 52, no. 2, 2016.
9. A. Roth, "Buying private data at auction: the sensitive surveyor's problem." *SIGecom Exchanges*, vol. 11, no. 1, pp. 1–8, 2012.
10. A. Ghosh and A. Roth, "Selling privacy at auction," in *Proceedings of the 12th ACM conference on Electronic commerce*. ACM, 2011, pp. 199–208.
11. L. K. Fleischer and Y.-H. Lyu, "Approximately optimal auctions for selling privacy when costs are correlated with data," in *Proceedings of the 13th ACM Conference on Electronic Commerce*. ACM, 2012, pp. 568–585.
12. K. Ligett and A. Roth, "Take it or leave it: Running a survey when privacy comes at a cost," in *Internet and Network Economics*. Springer, 2012, pp. 378–391.
13. K. Nissim, S. Vadhan, and D. Xiao, "Redrawing the boundaries on purchasing data from privacy-sensitive individuals," in *Proceedings of the 5th conference on Innovations in theoretical computer science*. ACM, 2014, pp. 411–422.
14. C. Dwork, "Differential privacy," in *Automata, languages and programming*. Springer, 2006, pp. 1–12.
15. J.-J. Laffont and D. Martimort, *The theory of incentives: the principal-agent model*. Princeton University Press, 2009.

16. L. Xu, C. Jiang, Y. Chen, Y. Ren, and K. J. R. Liu, "Privacy or utility in data collection? a contract theoretic approach," *IEEE Journal of Selected Topics in Signal Processing*, vol. 9, no. 7, pp. 1256–1269, Oct 2015.
17. D. Kirk, *Optimal Control Theory: An Introduction*, ser. Dover Books on Electrical Engineering. Dover Publications, 2012.
18. K. Bache and M. Lichman, "UCI machine learning repository," 2013. [Online]. Available: http://archive.ics.uci.edu/ml
19. K. LeFevre, D. J. DeWitt, and R. Ramakrishnan, "Incognito: Efficient full-domain k-anonymity," in *Proceedings of the 2005 ACM SIGMOD International Conference on Management of Data*, ser. SIGMOD '05.   New York, NY, USA: ACM, 2005, pp. 49–60. [Online]. Available: http://doi.acm.org/10.1145/1066157.1066164
20. F. Kohlmayer, F. Prasser, C. Eckert, A. Kemper, and K. Kuhn, "Flash: Efficient, stable and optimal k-anonymity," in *Privacy, Security, Risk and Trust (PASSAT), 2012 International Conference on and 2012 International Confernece on Social Computing (SocialCom)*, Sept 2012, pp. 708–717.

Chapter 4
Dynamic Privacy Pricing

Abstract Personal data market provides a promising way to deal with the conflict between exploiting the value of personal data and protecting individuals' privacy. However, determining the price of privacy is a tough issue. In this chapter, we study the pricing problem in a scenario where a data collector sequentially buys data from multiple data providers whose valuations of privacy are randomly drawn from an unknown distribution. To maximize the total payoff, the collector needs to dynamically adjust the prices offered to the providers. We model the sequential decision-making problem of the collector as a multi-armed bandit problem with each arm representing a candidate price. Specifically, the privacy protection technique adopted by the collector is taken into account. Protecting privacy generally causes a negative effect on the value of data, and this effect is embodied by the time-variant distributions of the rewards associated with arms. Based on the classic upper confidence bound policy, we propose two learning policies for the bandit problem. The first policy estimates the expected reward of a price by counting how many times the price has been accepted by data providers. The second policy treats the time-variant data value as a context and uses ridge regression to estimate the rewards in different contexts. Simulation results on real-world data demonstrate that by applying the proposed policies, the collector can get a payoff approximating to that he can get by setting a fixed price, which is the best in hindsight, for all data providers.

4.1 Introduction

In current wave of big data, the value of personal data has become more and more prominent. In the meantime, how to deal with the tension between exploiting the value of personal data and protecting individual privacy has become an important issue [1]. Researchers have made great effort in improving the data analysis techniques so as to avoid violating individuals' privacy [2–4]. Recently, the economic analysis of privacy [5] has been receiving growing attention. As a type of economic goods, personal data can be sold on the market by its owner. By this way, the industry demand for personal data can be satisfied, meanwhile, the data

© Springer International Publishing AG, part of Springer Nature 2018   89
L. Xu et al., *Data Privacy Games*, https://doi.org/10.1007/978-3-319-77965-2_4

owner can benefit from revealing some privacy. Selling personal data on markets is a plausible solution for the privacy-concerned individuals.

There are now some initiatives on personal data market where individuals can decide for themselves whether they allow the usage of their personal data [6]. One challenging issue for implementing personal data market is to determine the price of privacy. Different individuals generally have different definitions for privacy, and their attitudes towards privacy may be influenced by how the data are collected and who will use the data. Hence, the monetary value of privacy is highly context-dependent [5]. To complicate matters, even if the data provider, i.e. the individual, has a determined valuation of privacy, this valuation is generally unknown to others. In other words, there is *information asymmetry* between the data provider and the data collector who has a demand for personal data.

In the previous chapter, we have proposed a contract theoretical approach to deal with the information asymmetry problem. To design the optimal contract, it is assumed that the collector has a full knowledge of the distribution from which the data provider's privacy valuation is drawn. However, such an assumption is a bit idealistic. In this chapter, we study the privacy pricing problem in a setting where a data collector interacts with multiple data providers sequentially [8]. Each data provider has a data record that is desired by the collector. And the provider's valuation of the data is drawn from some probability distribution which is *unknown* to the collector. Similar to previous work [9, 10], the online posted-price mechanism is adopted. That is, each time a new data provider arrives, the collector offers the provider a price, and the provider will sell his data if and only if the price is higher than his valuation of the data. In order to maximize the total payoff obtained from the data collection process, the collector needs to adjust the price dynamically based on some learning policy. During the learning process, the collector faces a trade-off between staying with the price that has brought the highest payoff in the past and trying new prices that might bring higher payoffs in the future. A multi-armed bandit problem [11] is formalized to deal with the *exploitation-exploration* trade-off.

Different from current studies on bandit problems, we incorporate the idea of privacy protection into the dynamic pricing problem. Specifically, suppose that the collector applies anonymization techniques [3] to the collected data so as to protect data providers' privacy. The information loss caused by anonymization, which is related to the volume of the collected data, will decrease the value that the data brings to the collector. Considering that the data volume changes over time and depends on past interactions between the collector and data providers, we model the pricing problem as a multi-armed bandit problem with *time-variant* distributions of rewards. Based on the UCB (upper confidence bound) policy proposed for classic bandit problems, we develop learning polices to adapt to the time varying characteristic. The basic idea is to use the number of successful deals in the past and the latest estimate of the data value to compute the upper confidence bound of rewards. To evaluate the performance of the learning polices, we adopt the notion of *weak regret* introduced in [12]. The proposed policies are compared with a benchmark policy that always chooses the single globally best price. We conduct a series of simulations on real-world data. The simulation results show that the regrets of the proposed learning policies grow slowly as time evolves. And the price which is most frequently chosen by the policy is almost the same with the best single price that maximizes the total payoff of the collector.

The rest of the chapter is organized as follows. Section 4.2 briefly introduces some studies that are related to our work. The system model and the bandit formulation is presented in Sect. 4.3. Details of the learning polices proposed for the pricing problem are described in Sect. 4.4. In Sect. 4.5, we make a comparison of different learning polices and analyze the influence of parameters, based on the simulation results. Finally, this chapter is concluded in Sect. 4.6.

4.2 Related Work

4.2.1 Pricing Data

One important issue for successfully implementing personal data markets is to determine the monetary value of personal data. Many studies on this subject formalize the demand for personal data as queries over data. In [13], Li et al. proposed a framework for assigning prices to query answers which are perturbed to protect the privacy of data providers. In [14], Koutris et al. identified two important properties, namely arbitrage-free and discount-free, for the pricing function. Lin and Kifer [15] also studied the arbitrage-free property of query-based pricing, and they showed that for certain queries, the data seller has to accept some risk of arbitrage so as to set reasonable prices.

In contrast to these studies, some researchers proposed to directly assign prices to data. In [16], Gkatzelis et al. considered a market where buyers pay for access to unbiased samples of private data. A mechanism was proposed to incentivize individuals to truthfully report their privacy attitudes. In [17], Li and Raghunathan developed a pricing mechanism for the data provider to distribute sensitive data, where the purpose of data usage and different sensitivity levels of data were taken into consideration. We also study how to price individual's sensitive data. However, different from [17], we model the problem from the standpoint of the buyer (i.e. a data collector) rather than the seller (i.e. a data provider).

4.2.2 Dynamic Pricing and Bandit Problems

Dynamic pricing, which concerns optimally setting prices of products or services in a changeable market environment, usually gives rise to a trade-off between maximizing the instant reward and learning unknown properties of the environment [18]. This exploitation-exploration trade-off is widely investigated in the reinforcement learning area [19–21], especially in the study of multi-armed bandit problems [11].

In the literature on dynamic pricing, bandit problems are often applied to "posted price" models where the seller provides the buyer a take-it-or-leave-it price offer. For example, by exploiting the link between procurement auctions and multi-armed bandits, Singla and Krause [9] proposed a posted-price mechanism which is budget feasible and incentive compatible. In [10], Amin et al. considered a scenario where a seller repeatedly interacts with a strategic buyer who wants to maximize his long-term surplus. They introduced the definition of strategic regret and developed learning algorithms that are no-regret with respect to this definition.

In the study of classic stochastic multi-armed bandit problems, it is assumed that the reward distributions of arms do not change over time. While in some cases, temporal changes of the reward distribution is intrinsic to the problem [22]. In [23], Vakili et al. studied time-varying bandit problems where reward distribution can change arbitrarily over time. In [24], Garivier and Moulines also showed that the learning policies proposed for classic bandit problems can be adapted to cope with the non-stationary cases. Different from above studies where the change of reward distribution is arbitrary, in our case, how the reward distribution changes heavily depends on the past choices of arms. Therefore, rather than directly applying previously proposed learning policies, we need to develop new policies for our problem.

4.3 System Model and Problem Formulation

4.3.1 Privacy Pricing

Consider a scenario where one data collector (e.g. a website) interacts with multiple individuals. Each individual, referred to as a *data provider*, owns a data record which is desired by the collector. If the data provider gives his data to the collector, he may suffer a loss because of the disclosure of privacy. To compensate the privacy loss, the collector pays monetary rewards to the data provider. Let p denote the price that the collector is willing to pay for one data record.

Upon receiving the collector's price offer, the data provider can decide whether or not to provide his data. Different data providers have different attitudes towards privacy, thus they show different responses to the same price offer. The privacy attitude of a data provider is quantified by a parameter θ. A large θ means the provider cares much about privacy. Moreover, we assume θ is drawn independently and identically from [0, 1]. The corresponding probability density function is denoted by $f(\theta)$. In following descriptions, we sometimes refer to θ as the *type* of the data provider. The parameter θ can also be interpreted as the cost that the data provider pays for producing one data record. In other words, the data provider will provide his data to the collector if and only if $p \geq \theta$.

To protect data providers' privacy, the collector applies some anonymization technique to the collected data. After the anonymization process, the collector can conduct data mining or sell the data to a third party. Either way, the collector derives values from the data. Let v denote the value that one data record brings to the collector. Then the payoff that the collector obtains from the deal with one data provider is

$$u\left(p;\theta\right) = \begin{cases} v-p, & p \geq \theta \\ 0, & 0 \leq p < \theta \end{cases}. \qquad (4.1)$$

The payoff $u\left(p;\theta\right)$ can be seen as a parametric function with θ being the parameter. Given the distribution of θ and the price p, the expected payoff can be written as

$$\mathbb{E}\left[u\left(p;\theta\right)\right] = F_\theta\left(p\right)\left(v-p\right), \qquad (4.2)$$

where $F_\theta\left(p\right) \triangleq \int_0^p f\left(\theta\right)d\theta$ denotes the probability that a data provider accepts the price p.

The goal of the collector is to set a proper price to maximize the expected payoff. However, the data provider's θ is usually unknown to the collector, which makes it difficult for the collector to make the optimal decision. In our previous work [7], a contract theoretical approach was proposed to find the optimal pricing rule. The contract approach was based on the assumption that the distribution $f\left(\theta\right)$ is known to the collector. While in this paper, we consider a more practical setting where the collector has no preliminary knowledge about the distribution apart from the fact that its support is in [0, 1]. In such a case, the collector cannot directly determine the optimal price but has to learn it gradually through multiple interactions with data providers.

4.3.2   Bandit Formulation

When interacting with data providers sequentially, the collector faces a trade-off between *exploiting* current knowledge to focus on the price that has brought the highest payoff so far and *exploring* new prices that might bring higher payoff in the future. Designing a learning policy to solve the exploitation-exploration trade-off is usually formalized as a bandit problem. Different from previous studies on bandit problems, we are more interested in how to embody the concept of privacy in the bandit formulation. Next we first introduce some basics of the stochastic bandit problem, then we discuss how to formalize the influence of privacy protection on data collector's payoff.

Consider a data collection scenario where time evolves in rounds. At each round $t \in \{1, 2, \cdots, T\}$, a new data provider with type θ_t arrives. The collector chooses a price from the set $P \triangleq \left\{p_i | p_i = \frac{i}{K}, i = 1, \cdots, K\right\}$ and offers it to the data provider. Here we restrict the price to a set of discrete values for simplicity reasons. Though the collector may suffer a loss due to the discretization, the loss becomes smaller as the price range is discretized more finely. Following the bandit terminology, each price $p_i \in P$ is called an *arm*. If the collector chooses p_i for data provider of type θ_t, the collector can get a reward $r_{i,t} \triangleq u(p_i; \theta_t)$. Considering that θ_t is randomly distributed within $[0, 1]$, the reward is randomly drawn from some unknown probability distribution that is associated with the arm. For each arm $p_i \in P$, the expected reward is

$$\mu_i = F_\theta(p_i)(v - p_i). \tag{4.3}$$

Without the prior knowledge of $F_\theta(\cdot)$, the collector applies a learning policy to find the best arm p_{I^*}, where $I^* = \arg\max_{i=1,\cdots,K} \mu_i$.

A learning policy can be formalized as a set of maps $\{\sigma_t\}$ where σ_t is a map from the observed history up to round $t - 1$ to the index of the arm to be chosen at round t, denoted as I_t. The performance of the learning policy is evaluated by *regret* [11], which is the difference between the rewards accumulated by the policy and the rewards accumulated by a hypothetical benchmark policy that always chooses the best arm. Since both the rewards and the choices of arms are stochastic, the regret is usually computed as follows:

$$R(T) = T\mu_{I^*} - \mathbb{E} \sum_{t=1}^{T} \mu_{I_t}, \tag{4.4}$$

where T is the given time horizon, and the expectation is taken over the possible randomness of the learning policy.

4.3.3 Arms with Time-Variant Rewards

The stochastic bandit formulation introduced above has an implicit assumption that each arm is associated with a time-invariant distribution of reward, thereof the best arm remains unchanged over time. However, this assumption fails to hold in our problem setting. As mentioned earlier, before making use of the collected data records, the collector performs data anonymization to protect data providers' privacy. Suppose that the collector chooses k-anonymity [25] as the privacy criterion. A data set is called k-anonymous if each record is indistinguishable from at least $k - 1$ other records with respect to quasi-identifiers. Table 4.1 shows an example of 2-anonymity. Roughly speaking, the probability that one individual being identified from the k-anonymous data set is less than $\frac{1}{k}$. A large k indicates a high level of privacy security. Here we assume the value of k is pre-specified by the collector and will not be changed during the data collection process.

A common approach to realize k-anonymity is to generalize the data. Such an operation causes a decline in data utility. In other words, information loss is inevitable. One thing we want to point out is that as the collector gathers

Table 4.1 An example of 2-anonymity, where quasi-identifiers are *Age*, *Gender*, and *Zipcode*

Age	Gender	Zipcode	Disease
(a) Original table			
5	Female	12000	HIV
9	Male	14000	Dyspepsia
6	Male	16000	Dyspepsia
8	Female	19000	Bronchitis
12	Female	21000	HIV
15	Female	22000	Cancer
17	Female	26000	Pneumonia
19	Male	27000	Gastritis
21	Female	33000	Flu
24	Male	37000	Pneumonia
(b) 2-anonymous table			
[1, 10]	People	1****	HIV
[1, 10]	People	1****	Dyspepsia
[1, 10]	People	1****	Dyspepsia
[1, 10]	People	1****	Bronchitis
[11, 20]	People	2****	HIV
[11, 20]	People	2****	Cancer
[11, 20]	People	2****	Pneumonia
[11, 20]	People	2****	Gastritis
[21, 60]	People	3****	Flu
[21, 60]	People	3****	Pneumonia

more data records, a smaller degree of generalization would be enough to meet the specified privacy criterion, and correspondingly, the information loss becomes smaller. Consider the following example. Suppose that the collector gets the first 3 records in Table 4.1a during the collection process. To realize 2-anonymity, values of all the three quasi-identifiers, i.e. *Age*, *Gender* and *Zipcode*, need to be generalized. However, if the collector also gets the fourth record, then there is no need to generalize *Gender*. We have conducted simulations on real-world data to examine the relationship between the information loss and the value of k. Details will be presented in Sect. 4.5.2.

Now let's back to the bandit formulation. As defined in (4.3), given the distribution of θ, the expected reward of each arm is determined by v which is the value that one anonymized data record brings to the collector. Let v_{ori} denote the value of the original data record. Due to the information loss caused by anonymization, there is $v < v_{ori}$. According to above discussion, the difference between v and v_{ori} depends on the size of the collected data set. Let N_t denote the number of data records that the collector has obtained by the end of round t. If the collector performs anonymization on a data set of size N_t, the value of the anonymized data record can be defined as

$$v_t = \begin{cases} v_{\min}, \ if \ N_t < k, \\ (1 - \rho\,(N_t;k))\,v_{ori}, \ otherwise, \end{cases} \qquad (4.5)$$

where $v_{\min}$ is a small positive constant, meaning that the data record is of little value to the collector when the data size is too small, since the collector cannot use the data without compromising the required privacy criterion. The parameterized function $\rho\,(\cdot;k)$ computes the average information loss caused by k-anonymity. It is decreasing, or at least non-increasing, with N_t. And $\rho\,(\cdot;k)$ is known to the collector.

Given the time horizon T, the collector will get a data set of size N_T at the end of the data collection process, and each data record brings value v_T to the collector. Therefore, in hindsight, the expected reward of the arm p_i is

$$\mu_i = F_\theta\,(p_i)\,(v_T - p_i)\,. \qquad (4.6)$$

In above equation, the precise value of v_T is unknown to the collector until the collection process stops. At each round $t \in \{1, \cdots, T\}$, the collector can only predict the value based on current data size N_t. Suppose that the collector always makes a conservative estimation. That is, the collector uses v_t to evaluate the rewards. Define the expected reward of arm p_i at round t as

$$\mu_{i,t} = F_\theta\,(p_i)\,(v_t - p_i)\,. \qquad (4.7)$$

According to (4.5) and (4.7), the expected reward of each arm varies with time. It should be noted that though the variation of v_t is the same for all arms, the variation of the expected reward is arm-specific, which means the best arm is not fixed. More specifically, let ε_t denote the increase of data value from round $t - 1$ to round t, i.e. $v_t = v_{t-1} + \varepsilon_t$. For each arm p_i, the corresponding increase in expected reward is

$$\Delta\mu_{i,t} = \mu_{i,t} - \mu_{i,t-1} = F_\theta\,(p_i)\,\varepsilon_t\,. \qquad (4.8)$$

Above equation implies that different arms have different growth rates of expected reward. As a result, the best arm of current round may fall behind some other arm in next round.

The fact that the distributions of rewards are time-variant differs the privacy pricing problem from classic stochastic bandit problems. At a high level, our problem resembles the adversarial bandit problems [11, 12] which makes no statistical assumptions about the generation of rewards. The adversarial bandit problem assumes that the sequence of rewards of each arm is specified by an adversary. In our case, considering that the rewards of arms at each round are heavily dependent on the collector's historical choices, we can assume there is a *non-oblivious* adversary who specifies the rewards based on the collector's past behavior. However, analyzing such a bandit problem is intricate. Realizing that it is still possible to use stochastic distributions, though not stationary distributions, to model the rewards of arms, we choose to adapt the learning policies proposed for stochastic bandit problems to our problem. Nevertheless, the regret defined in (4.4) is no longer appropriate for the time-invariant case. Instead, we use the notion of *weak regret* [12], which is proposed for adversarial bandit problems, to measure the performance of the learning policy. Given the time horizon T, the weak regret is defined as

$$R\left(T\right) = \max_{i=1,\cdots,K} \sum_{t=1}^{T} r_{i,t} - \sum_{t=1}^{T} r_{I_t,t}. \tag{4.9}$$

Above equation means that the learning policy is competed against the one that constantly chooses the single globally best arm.

4.4 Learning Policy

As a standard tool from statistics, *confidence bound* [26] is commonly used to deal with the exploitation-exploration trade-off in bandit problems. Based on the basic UCB (upper confidence bound) policy, we propose two approaches to deal with the time-variant issue described in above section. The first approach directly modifies the basic UCB policy to make a more accurate estimate of the expected reward. The second approach treats the time-variant data value as a context and adopts a contextual bandit model to formulate the problem.

4.4.1 Upper Confidence Bound

The learning policy UCB1 proposed in [27] and its variants are widely applied to bandit problems. The basic idea of UCB1 is to estimate the unknown expected reward of each arm by making a linear combination of previously observed rewards of the arm. During the learning procedure, the policy maintains two quantities, namely n_i and $\bar{r}_i$, for each arm p_i. The first quantity $n_i \triangleq \sum_{\tau=1}^{t-1} \mathbb{1}\left(I_\tau = i\right)$ denotes how many times that p_i has been chosen up to round t. The second quantity $\bar{r}_i$ is the average of the rewards observed for p_i, and $\bar{r}_i$ is treated as an estimate of the true expected reward with $\bar{r}_i + \alpha\sqrt{\frac{\ln t}{n_i}}$ being the upper confidence bound, where the parameter α controls the width of the confidence interval. At each round, the arm which currently has the maximal upper confidence bound of reward is chosen. If we ignore the variance of v_t across time rounds, we can directly apply this UCB policy to the privacy pricing problem. A detailed description of the policy is shown in Algorithm 1.

Algorithm 1 UCB

Require: $\alpha \in \mathbb{R}^+$
 1: **for** $t = 1$ to K **do**
 2:     Choose arm $I_t = t$
 3:     Observe and record the result $\mathbb{1}\left(p_{I_t} \geq \theta_t\right)$
 4:     $N_t = \begin{cases} N_{t-1} + 1, & if \ p_{I_t} \geq \theta_t \\ N_{t-1}, & otherwise \end{cases}$
 5:     Compute v_t based on N_t
 6:     $r_t \leftarrow \left(v_t - p_{I_t}\right) \mathbb{1}\left(p_{I_t} \geq \theta_t\right)$
 7:     $n_{I_t} \leftarrow 1$
 8: **end for**
 9: **for** $t = K + 1$ to T **do**
10:     **for** $i = 1$ to K **do**
11:         $\bar{r}_i \leftarrow \frac{1}{n_i} \sum_{\tau=1}^{t-1} r_\tau \mathbb{1}\left(I_\tau = i\right)$
12:     **end for**
13:     Choose arm $I_t = \underset{i=1,\cdots,K}{\mathrm{argmax}} \left(\bar{r}_i + \alpha \sqrt{\frac{\ln t}{n_i}}\right)$ with ties broken arbitrarily
14:     Observe and record the result $\mathbb{1}\left(p_{I_t} \geq \theta_t\right)$
15:     $N_t = \begin{cases} N_{t-1} + 1, & if \ p_{I_t} \geq \theta_t \\ N_{t-1}, & otherwise \end{cases}$
16:     Compute v_t based on N_t
17:     $r_t \leftarrow \left(v_t - p_{I_t}\right) \mathbb{1}\left(p_{I_t} \geq \theta_t\right)$
18:     $n_{I_t} \leftarrow n_{I_t} + 1$
19: **end for**

Informally, if the arm with large $\alpha\sqrt{\frac{\ln t}{n_i}}$ is chosen, we can say that the collector makes an explorative decision, since in such a case taking $\bar{r}_i$ as the estimate of the true expected reward is quite unreliable. Contrarily, if an arm with large $\bar{r}_i$ is chosen, we say the collector makes an exploitative decision. Considering that $\alpha\sqrt{\frac{\ln t}{n_i}}$ decreases rapidly with each choice of p_i, the number of explorative decisions is limited. As $\alpha\sqrt{\frac{\ln t}{n_i}}$ becomes smaller, the average $\bar{r}_i$ gets closers to the true expected reward, and it is more likely that the arm corresponding to maximal $\bar{r}_i$ is indeed the best arm.

4.4.2 Estimating Cumulative Distribution

The policy UCB described above takes the average of historical rewards as the estimate of the expected reward of an arm. However, when the reward varies with time, as we mentioned in Sect. 4.3.3, such estimation is inappropriate. From (4.7) we can see that, at each round t, the collector only needs to know the value of $F_\theta(p_i)$ to estimate the expected reward, since the collector can compute the value of v_t based on current data size N_t. The cumulative distribution $F_\theta(\cdot)$ determines the success probability of each price. By "success" we mean the price offer is accepted by the data provider. Thus, we can estimate $F_\theta (p_i)$ by simply counting how many times that p_i is chosen by the collector and accepted by data providers. Specifically, at round t, the estimate of $F_\theta (p_i)$ is given by

$$\hat{F}_\theta (p_i) = \frac{\sum_{\tau=1}^{t-1} \mathbb{1}(I_\tau = i) \mathbb{1}(p_i \geq \theta_\tau)}{n_i}, \qquad (4.10)$$

where n_i has the same meaning as we defined before, and $\mathbb{1}(p_i \geq \theta_\tau)$ indicates whether p_i was accepted by the data provider who arrived at round τ.

Based on above discussion, we propose the following approach to modify the policy UCB to make a more accurate estimate of the expected reward. Initially, similar to UCB, each arm is chosen once, and the corresponding result $\mathbb{1}(p_i \geq \theta_\tau)$ is recorded. Then at the beginning of each round t, the collector observes N_{t-1} and computes v_{t-1}. Based on v_{t-1} and historical information $\left\{(I_\tau, \mathbb{1}(p_{I_\tau} \geq \theta_\tau))\right\}_{\tau=1}^{t-1}$, the collector estimates the expected reward for each arm and then chooses the arm corresponding to the maximal upper confidence bound. Algorithm 2 gives a detailed description of this learning policy. In subsequent discussions, we refer to this modified policy as *VarUCB*.

Algorithm 2 VarUCB

Require: $\alpha \in \mathbb{R}^+$
1: **for** $t = 1$ to K **do**
2:      Choose arm $I_t = t$
3:      Observe and record the result $\mathbb{1}(p_{I_t} \geq \theta_t)$
4:      $N_t = \begin{cases} N_{t-1} + 1, & if\ p_{I_t} \geq \theta_t \\ N_{t-1}, & otherwise \end{cases}$
5:      Compute v_t based on N_t
6:      $r_t \leftarrow (v_t - p_{I_t}) \mathbb{1}(p_{I_t} \geq \theta_t)$
7:      $n_{I_t} \leftarrow 1$
8: **end for**
9: **for** $t = K + 1$ to T **do**
10:     Observe N_{t-1} and compute v_{t-1} by using (4.5)
11:     **for** $i = 1$ to K **do**
12:         $\hat{F}_\theta (p_i) = \frac{\sum_{\tau=1}^{t-1} \mathbb{1}(I_\tau = i) \mathbb{1}(p_i \geq \theta_\tau)}{n_i}$
13:     **end for**
14:     Choose arm $I_t = \underset{i=1,\cdots,K}{\operatorname{argmax}} \left(\hat{F}_\theta (p_i) (v_{t-1} - p_i) + \alpha \sqrt{\frac{\ln t}{n_i}} \right)$ with ties broken arbitrarily
15:     Observe and record the result $\mathbb{1}(p_{I_t} \geq \theta_t)$
16:     $N_t = \begin{cases} N_{t-1} + 1, & if\ p_{I_t} \geq \theta_t \\ N_{t-1}, & otherwise \end{cases}$
17:     $n_{I_t} \leftarrow n_{I_t} + 1$
18: **end for**

4.4.3 Contextual Bandit Approach

Instead of estimating the cumulative distribution, here we view the time-variant characteristic of the bandit problem from a different perspective. As discussed in Sect. 4.3.2, at each round t, the reward distribution of each arm is determined by current value of the anonymized data record. And the data value mainly depends on how many data that the collector has already got. The number of data records N_t can be seen as a type of side information associated with each arm. Moreover, if we treat this side information as the *context*, then the pricing problem can be cast into a contextual bandit problem [11, 28]. A formal description of this contextual bandit problem is given below.

At the beginning of round t, the collector observes the context N_{t-1}. Based on the context, the collector computes v_{t-1} by using (4.5). We define $\mathbf{x}_t = (v_{t-1}, 1)^T$ as the *feature vector* representing the context. Then the expected reward of arm p_i can be expressed as

$$\mu_{i,t} = \mathbf{x}_t^T \boldsymbol{\omega}_i^*, \tag{4.11}$$

where $\boldsymbol{\omega}_i^* \triangleq (F_\theta(p_i), -p_i F_\theta(p_i))^T$ represents the unknown coefficient vector.

To determine the best arm of each round, the collector needs to learn the optimal mapping of contexts to arms. A key step in the learning procedure is to estimate the expected rewards of arms. The formulation shown in (4.11) fits the basic form of linear regression: features of the context are independent variables, and the expected reward is the dependent variable. Therefore, we can treat the observed context-reward pairs as training samples and train a regression model for each arm. Specifically, let $\tau_1^i, \tau_2^i, \cdots, \tau_{n_i}^i$ denote the sequence of rounds at which p_i is chosen. Let $\mathbf{D}_i \in \mathbb{R}^{m \times 2}$ denote the m contexts that are recently observed for p_i, i.e. $\mathbf{D}_i = \left[\mathbf{x}_{\tau_{n_i-(m-1)}^i} \quad \mathbf{x}_{\tau_{n_i-(m-2)}^i} \quad \cdots \mathbf{x}_{\tau_{n_i}^i}\right]^T$. Let $\mathbf{c_i} \in \mathbb{R}^m$ be the vector of observed rewards corresponding to these contexts.

Given the training data $(\mathbf{D}_i, \mathbf{c}_i)$, we can make a least square estimation of the coefficient vector $\boldsymbol{\omega}_i^*$, which is $\widehat{\boldsymbol{\omega}}_i = (\mathbf{D}_i^T \mathbf{D}_i)^{-1} \mathbf{D}_i^T \mathbf{c}_i$. This is the most common way to solve the linear regression problem. An important premise for the least square estimation to work is that the matrix $\mathbf{D}_i^T \mathbf{D}_i$ cannot be singular or nearly singular. By simple computation we get

$$\det\left(\mathbf{D}_i^T \mathbf{D}_i\right) = m \sum_{j=0}^{m-1} v_{\tau_{n_i-j}^i}^2 - \left(\sum_{j=0}^{m-1} v_{\tau_{n_i-j}^i}\right)^2. \tag{4.12}$$

Considering that the variation of v_t across different rounds may be very small, it is likely that $\det(\mathbf{D}_i^T \mathbf{D}_i)$ approaches zero. For example, if we set $m = 2$, then

$\det\left(\mathbf{D}_i^T \mathbf{D}_i\right) = \left(v_{\tau_{n_i}^i} - v_{\tau_{n_i-1}^i}\right)^2$, which implies the matrix $\mathbf{D}_i^T \mathbf{D}_i$ will become singular when $v_{\tau_{n_i}^i} = v_{\tau_{n_i-1}^i}$.

Inspired by Li et al.'s work [28], we apply *ridge regression* to overcome the shortcoming of least square estimation. The estimate of $\boldsymbol{\omega}_i^*$ is given by

$$\widehat{\boldsymbol{\omega}}_i = \left(\mathbf{D}_i^T \mathbf{D}_i + \mathbf{I}\right)^{-1} \mathbf{D}_i^T \mathbf{c}_i, \qquad (4.13)$$

where $\mathbf{I}$ is the 2×2 identity matrix.

Theorem 4.1 *Let* $\mathbf{A}_i = \mathbf{D}_i^T \mathbf{D}_i + \mathbf{I}$ *and* $\alpha = 1 + \sqrt{\frac{1}{2}\ln\frac{2T}{\eta}}$, *where* η *is a positive constant and* $\eta < T$. *When components in* $\mathbf{c}_i$ *are independent conditioned on corresponding rows in* $\mathbf{D}_i$, *then with probability at least* $1 - \frac{\eta}{T}$, *there is*

$$\left|\mathbf{x}_t^T \widehat{\boldsymbol{\omega}}_i - \mathbf{x}_t^T \boldsymbol{\omega}_i^*\right| \leq \alpha \sqrt{\mathbf{x}_t^T \mathbf{A}_i^{-1} \mathbf{x}_t}. \qquad (4.14)$$

Proof Please see the appendix for the proof.

The product of $\mathbf{x}_t$ and $\widehat{\boldsymbol{\omega}}_i$ gives an estimate of the true expected reward. According to (4.14), the estimate is upper bounded by $\mathbf{x}_t^T \widehat{\boldsymbol{\omega}}_i + \alpha \sqrt{\mathbf{x}_t^T \mathbf{A}_i^{-1} \mathbf{x}_t}$. With this confidence bound, a UCB-based learning policy can be derived. A detailed description of the policy is given in Algorithm 3. Similar to [28], we refer to this policy as LinUCB.

Algorithm 3 LinUCB
___
Require: $\alpha \in \mathbb{R}^+$
1: **for** $t = 1$ to T **do**
2:     Observe features of current context $\mathbf{x}_t = (v_{t-1}, 1)^T$
3:     **for** $i = 1$ to K **do**
4:         **if** p_i is new **then**
5:             $\mathbf{A}_i \leftarrow \mathbf{I}$ (2-dimensional identity matrix)
6:             $\mathbf{b}_i \leftarrow \mathbf{0}$ (2-dimensional zero vector)
7:         **end if**
8:         $\widehat{\boldsymbol{\omega}}_i \leftarrow \mathbf{A}_i^{-1} \mathbf{b}_i$
9:         $\widehat{\mu}_{i,t} \leftarrow \mathbf{x}_t^T \widehat{\boldsymbol{\omega}}_i + \alpha \sqrt{\mathbf{x}_t^T \mathbf{A}_i^{-1} \mathbf{x}_t}$
10:     **end for**
11:     Choose arm $I_t = \underset{i=1,\cdots,K}{\operatorname{argmax}} \mu_{i,t}$ with ties broken arbitrarily
12:     Observe the result $\mathbb{1}\left(p_{I_t} \geq \theta_t\right)$
13:     $N_t = \begin{cases} N_{t-1} + 1, & \text{if } p_{I_t} \geq \theta_t \\ N_{t-1}, & otherwise \end{cases}$
14:     Compute v_t based on N_t
15:     $r_t = \left(v_t - p_{I_t}\right) \mathbb{1}\left(p_{I_t} \geq \theta_t\right)$
16:     $\mathbf{A}_{I_t} \leftarrow \mathbf{A}_{I_t} + \mathbf{x}_t \mathbf{x}_t^T$
17:     $\mathbf{b}_{I_t} \leftarrow \mathbf{b}_{I_t} + r_t \mathbf{x}_t$
18: **end for**

As described in Algorithm 3, the policy uses a matrix $\mathbf{A}_i$ to record the history of the context and a vector $\mathbf{b}_i$ to record the accumulative reward for each arm. More specifically, at the end of round t,

$$
\mathbf{A}_i = \begin{bmatrix} 1 + \sum_{\tau=1}^{t} v_{\tau-1}^2 \mathbb{1}\,(I_\tau = i) & \sum_{\tau=1}^{t} v_{\tau-1}^2 \mathbb{1}\,(I_\tau = i) \\ \sum_{\tau=1}^{t} v_\tau \mathbb{1}\,(I_\tau = i) & 1 + \sum_{\tau=1}^{t} \mathbb{1}\,(I_\tau = i) \end{bmatrix},
\tag{4.15}
$$

$$
\mathbf{b}_i = \begin{bmatrix} \sum_{\tau=1}^{t} (v_\tau - p_i)\, v_{\tau-1} \mathbb{1}\,(I_\tau = i)\, \mathbb{1}\,(p_i \geq \theta_\tau) \\ \sum_{\tau=1}^{t} (v_\tau - p_i)\, \mathbb{1}\,(I_\tau = i)\, \mathbb{1}\,(p_i \geq \theta_\tau) \end{bmatrix}.
\tag{4.16}
$$

As mentioned in Sect. 4.3.3, the true value of the anonymized data can only be determined after the whole data collection process stops. In other words, if the collection process stops after round T, then for any $t \in \{1, \cdots, T-1\}$, v_t actually represents the collector's estimate of the true value v_T. Considering this, we propose a new way to update $\mathbf{A}_i$ and $\mathbf{b}_i$ so that they can be consistent with the collector's latest estimate of v_T. The proposed update rule is quite intuitive, that is, we replace all the v_τ ($\tau = 1, \cdots, t-1$) in (4.15) and (4.16) with v_t which is computed based on current number of data records. Algorithm 4 gives a detailed description of the modified policy. In subsequent discussions, we refer to this learning policy as VarLinUCB.

4.5   Simulation

In the previous section we have proposed several learning policies for the dynamic privacy pricing problem. To evaluate the performance of the learning policies, we conduct simulations on real-world data. In the following part, we first introduce the dataset and the anonymization method. Then we present experiment results to demonstrate the relationship between the information loss and the number of data records. After that, we describe the experiment settings of learning polices and the evaluation method. Based on the simulation results, comparisons of different polices are made, and the influence of the parameters is analyzed.

4.5.1   Dataset and Anonymization Method

Simulations are conducted on the Adult data set [29], which is widely used in the study of data anonymization. The original data set consists of 32,561 records from a census database, and each record consists of 15 attributes. After removing records

---

Algorithm 4 VarLinUCB

Require: $\alpha \in \mathbb{R}^+$

1: **for** $t = 1$ to T **do**
2:     Observe features of current context $\mathbf{x}_t = (v_{t-1}, 1)^T$
3:     **for** $i = 1$ to K **do**
4:         **if** p_i is new **then**
5:             $\mathbf{A}_i \leftarrow \mathbf{I}$ (2-dimensional identity matrix)
6:             $\mathbf{b}_i \leftarrow \mathbf{0}$ (2-dimensional zero vector)
7:         **end if**
8:         $\widehat{\omega}_i \leftarrow \mathbf{A}_i^{-1}\mathbf{b}_i$
9:         $\widehat{\mu}_{i,t} \leftarrow \mathbf{x}_t^T \widehat{\omega}_i + \alpha \sqrt{\mathbf{x}_t^T \mathbf{A}_i^{-1} \mathbf{x}_t}$
10:     **end for**
11:     Choose arm $I_t = \underset{i=1,\cdots,K}{\mathrm{argmax}}\ \mu_{i,t}$ with ties broken arbitrarily
12:     Observe the result $\mathbb{1}\left(p_{I_t} \geq \theta_t\right)$
13:     $N_t = \begin{cases} N_{t-1} + 1, & if\ p_{I_t} \geq \theta_t \\ N_{t-1}, & otherwise \end{cases}$
14:     Compute v_t based on N_t
15:     $r_t = \left(v_t - p_{I_t}\right) \mathbb{1}\left(p_{I_t} \geq \theta_t\right)$
16:     $\mathbf{A}_{I_t} \leftarrow \begin{bmatrix} 1 + v_t^2 \sum_{\tau=1}^{t} \mathbb{1}\left(I_\tau = I_t\right) & v_t^2 \sum_{\tau=1}^{t} \mathbb{1}\left(I_\tau = I_t\right) \\ v_t \sum_{\tau=1}^{t} \mathbb{1}\left(I_\tau = I_t\right) & 1 + \sum_{\tau=1}^{t} \mathbb{1}\left(I_\tau = I_t\right) \end{bmatrix}$
17:     $\mathbf{b}_{I_t} \leftarrow \begin{bmatrix} \left(v_t - p_{I_t}\right) v_t \sum_{\tau=1}^{t} \mathbb{1}\left(I_\tau = I_t\right) \mathbb{1}\left(p_{I_t} \geq \theta_\tau\right) \\ \left(v_t - p_{I_t}\right) \sum_{\tau=1}^{t} \mathbb{1}\left(I_\tau = I_t\right) \mathbb{1}\left(p_{I_t} \geq \theta_\tau\right) \end{bmatrix}$
18: **end for**

---

with missing values, we use the remained 30,162 records for experiments. Similar to previous studies [30], nine attributes, including *age, workclass, education, marital-status, occupation, race, sex, native-country*, and *salary-class*, are chosen as quasi-identifiers.

To perform anonymization, we develop a java program based on the open source anonymization framework ARX [31], which supports different types of privacy criteria. Here we choose k-anonymity as the privacy criterion. The information loss is evaluated by the recommended default measure *Loss* [32], which summarizes the coverage of the domain of an attribute. The value of *Loss* ranges from 0 to 1. Large value indicates large information loss.

4.5.2 Relationship Between Information Loss and Data Size

As described in Sect. 4.3.1, a basic assumption of our study is that given the privacy criterion, the information loss caused by anonymization decreases as the number of

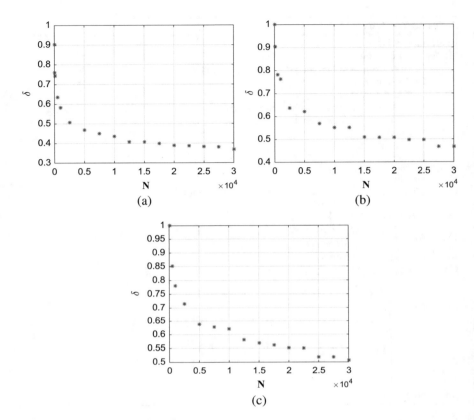

Fig. 4.1 Relationship between the information loss caused by k-anonymity and the number of data records. (**a**) $k = 10$. (**b**) $k = 50$. (**c**) $k = 100$

data records increases, thus the value of anonymized data increases over time. To validate the rationality of this assumption, we conduct a group of anonymization experiments on aforementioned data set. By randomly selecting N records from the data set, we construct 15 data sets of different sizes. Then for each $k \in \{10, 50, 100\}$, we run the anonymization program on these data sets respectively and record the corresponding information loss δ. From the experiment results shown in Fig. 4.1 we can see that, as the size of data set increases, the information loss decreases, and the decrease speed becomes slow as the data size becomes large. The results confirm our assumption.

In subsequent simulations of learning policies, we set $k = 10$ for the collector. To get a quantitative relation between the information loss and the number of data records, we use the curve fitting toolbox provided in MATLAB and formulate δ as a power function of N. Given $k = 10$, the parameterized function $\rho(N_t; k)$ in (4.5) is now defined as

$$\rho(N_t; k) = 1.193(N_t)^{-0.1104}. \tag{4.17}$$

Fig. 4.2 Relationship between the information loss caused by 10-anonymity and the number of data records. Red stars represent actual experiment results. The blue curve denotes the function $\rho\,(N_t;10) = 1.193(N_t)^{-0.1104}$ which is obtained by using MATLAB curve fitting toolbox. The reported R-square index is 0.9901, which indicates the fitting model is fine

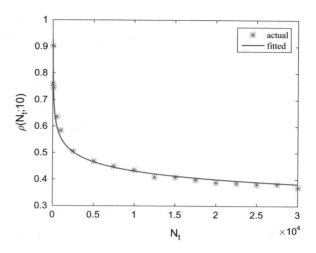

This fitting function is depicted in Fig. 4.2. It should be noted that above equation is used by the learning policy to estimate the value of anonymized data. While when we evaluate the performance of the learning policy, we run the anonymization program and use the actual information loss to compute the data value. Details of the evaluation will explained later.

4.5.3 Parameter Setting of Learning Policies

To simulate the privacy pricing scenario, first we need to determine the type, i.e. the privacy cost, of each data provider. Here we use two methods to determine data providers' types $\{\theta_i\}_{i=1}^{30,162}$. The first method assigns a uniformly distributed random value $\theta_i \in [0, 1]$ to each data provider. The second method draws a value $\widetilde{\theta}_i$ according to a normal distribution $N\left(\mu_\theta, \sigma^2\right)$ with $\mu_\theta = 0.5$ and $\sigma = \frac{1}{6}$. Considering that $\widetilde{\theta}_i$ may fall outside [0, 1], we define θ_i as

$$\theta_i = \begin{cases} 0, \; if \; \widetilde{\theta}_i < 0 \\ 1, \; if \; \widetilde{\theta}_i > 1 \\ \widetilde{\theta}_i, \; otherwise \end{cases}. \tag{4.18}$$

It is assumed that the collector chooses price from the set $P \triangleq \{p_i | p_i = \frac{i}{K}, i = 1, \cdots, K\}$. We set $K = 10$ in all simulations. Other parameters used in the learning policies are set as follows:

- k: the parameter of k-anonymity is set to 10 in all simulations.
- v_{ori}: we test two values of v_{ori}, namely $v_{ori} = 2$ and $v_{ori} = 10$. Considering that the price range is [0, 1], by setting $v_{ori} = 2$ we can ensure that the

collector gets non-negative payoff even when the data loses 50% of its utility after anonymization. A large v_{ori} implies that a small change of information loss will cause significant variation in data value, and consequently, the expected reward of each price will be sensitive to the number of collected records.

- α: as described in Algorithms 1–4, the input parameter α controls the width of the confidence interval of the estimated expected reward. For each policy, we conduct a group simulations with $\alpha \in \{0.1, 0.2, \cdots, 1.5\}$ so as to approximately determine the best value of α. Different learning policies are compared on the basis of their best results.

4.5.4 Evaluation Method

In Sect. 4.4 we have proposed four learning policies, namely *UCB*, *VarUCB*, *LinUCB* and *VarLinUCB*. Here, for comparison purpose, we propose another three simple policies. Let p_t denote the price that the collector chooses at round t. The first policy, referred to as *FixHalf*, always sets the price to the midpoint of the price range, i.e. $p_t = 0.5$. The second policy, referred to as *Random*, randomly chooses a price from the set P at each round. The third policy, referred to as *HalfValue*, always sets the price as half of current data value, i.e. $p_t = 0.5v_t$.

The objective of the collector is to collect a large number of data records without paying too much. Thus we use the actual payoff to measure the performance of the learning policy. Suppose the data collection process stops after round t, the total payoff to the collector is

$$ U(t) = N_t v_t - \sum_{\tau=1}^{t} p_\tau \mathbb{1}(p_\tau \geq \theta_\tau), \qquad (4.19) $$

where v_t is computed based on the actual information loss that is obtained by performing anonymization on the data set of size N_t. The second term in the right hand side denotes the total price that the collector has paid.

Simulations are conducted on the Adult data set described above. The total number of data records is 30,162, so we set the time horizon $T = 30,162$ in all simulations. To investigate how the performance of the learning policy varies with time, we set a group of check points $C = \{500, 1000, 2500, 5000, 7500 \cdots, 27,500, 30,162\}$. During the simulation, the anonymization program is invoked at each check point $t \in C$. The reported information loss δ is used to compute the true value of the anonymized data record, which is $v_t = (1 - \delta) v_{ori}$. Then based on current N_t, we can use (4.19) to compute the collector's payoff $U(t)$.

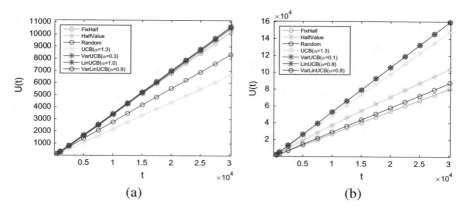

Fig. 4.3 Performance of different learning policies under the setting that data providers' types are uniformly distributed within [0, 1]. (**a**) $v_{ori} = 2$; (**b**) $v_{ori} = 10$

4.5.5   Simulation Results of Learning Policies

Given a group of $\{\theta_i\}_{i=1}^{30, 162}$ and a set of parameters $\{v_{ori}, \alpha\}$, we run each learning policy for ten times to reduce the influence of randomness, and the average result of each policy is reported.

4.5.5.1   Comparison of Different Policies

Figure 4.3 shows the simulation results of different policies where data providers' types are uniformly distributed. As we can see, the three variants of UCB, namely VarUCB, LinUCB, and VarLinUCB, have similar performance, and usually they are better than the other policies. When $v_{ori} = 2$, the simple policy HalfValue shows pretty good performance. This is because data providers' types are uniformly distributed within [0, 1], namely there is $F_\theta(p_t) = p_t$ ($0 \leq p_t \leq 1$). And by simple computation we know that, $p_t^* \triangleq 0.5v_t$ maximizes the expected reward defined in (4.7). The similar performance of VarUCB, LinUCB, VarLinUCB and HalfValue indicates that the three UCB-based policies successfully learn the distribution of data providers' types. Besides, since the difference between $0.5v_t$ and 0.5 is small, the policy FixHalf also performs well when $v_{ori} = 2$. However, when $v_{ori} = 10$, both HalfValue and FixHalf fail to produce an acceptable result. As we explained in Sect. 4.3.3, when evaluating the reward of an arm, the collector uses v_t as an estimate of v_T which is the true value of the anonymized data record. According to (4.5), the "error" of such estimation depends on both the information loss caused by anonymization and the original data value v_{ori}. The larger v_{ori} is, the less accurate the estimate is. Thus, when $v_{ori} = 10$, setting $p_t = 0.5v_t$ may bring the collector the maximal instant reward, while considering that the real reward is evaluated in

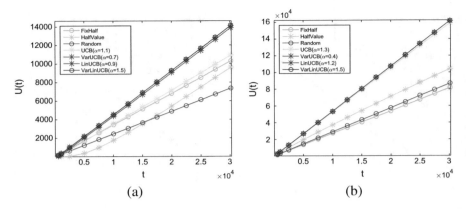

Fig. 4.4 Performance of different learning policies under the setting that data providers' types are approximately normally distributed within [0, 1]. (**a**) $v_{ori} = 2$; (**b**) $v_{ori} = 10$

hindsight, it is not a good choice. As for the policy FixHalf, the reason for its poor performance is simple. That is, when v_{ori} is large, even if v_t is very close to v_T, which means $0.5v_t$ can be approximately seen as the optimal price, setting $p_t = 0.5$ is quite different from setting $p_t = 0.5v_t$.

When data providers' types are normally distributed, as shown in Fig. 4.4, the three UCB-based policies, namely VarUCB, LinUCB, and VarLinUCB, still performance well. But the two simple policies, namely HalfValue and FixHalf, no longer show good performance, even when v_{ori} is small. This result is predictable, considering that the distribution of data provider's type has changed and $0.5v_t$ no longer maximizes the expected reward defined in (4.7). From Figs. 4.3b and 4.4b we can see that, when $v_{ori} = 10$, the performance gap between the policy UCB and the other three improved UCB policies is much smaller than that observed in the case of $v_{ori} = 2$. All these UCB policies utilize historical information to estimate the expect reward of an arm. The variation of reward's distribution, which is caused by the variation of v_t, is not taken into account by the policy UCB, thus UCB makes a worse estimate than the other three. However, when v_{ori} is large, the difference between v_t and v_T introduce a large error of the estimation, and such an error cannot be diminished by any of the policies. Therefore, in such a case, the advantages of the three improved policies become less obvious.

Above we have shown the simulation results on uniformly distributed data providers and normally distributed data providers. To further demonstrate the applicability of the proposed learning policies, we conduct following simulations. For each data provider, we draw a value θ_i from a power-low distribution.[1] We run each learning policy on this set of data owners. The rest parameters are set as before. Figure 4.5 shows the simulation results under the setting $v_{ori} = 2$. As

---

[1] http://tuvalu.santafe.edu/~aaronc/powerlaws/.

Fig. 4.5 Performance of different learning policies under the setting that $v_{ori} = 2$ and data providers' types follow the power-law distribution

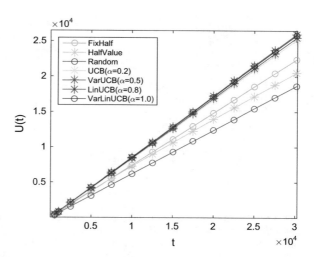

we can see, the three proposed learning policies, namely VarUCB, LinUCB, and VarLinUCB still perform better than others. Different from previous results, the basic UCB policy show good performance when the types are power-law distributed. The power-law distribution indicates most data providers have a small θ. During the learning procedure, the data collector frequently encounters a data provider with small θ. Suppose at the early rounds, a low price is chosen by the collector and it is accepted by such a data provider. Then the collector can stick to the choice in subsequent rounds, since it is very likely that the price will be accepted by latter users. In other words, simply based on the prices chosen in the past, the collector can make a proper decision in the future. This may explain why the basic UCB policy leads to a good result.

4.5.5.2 Weak Regret

To further demonstrate the performance of the proposed learning policies, we compute the weak regrets for the three UCB policies (i.e. VarUCB, LinUCB and VarLinUCB) and observe how the regrets change as time evolves. According to (4.9), in order to measure the weak regret, we need to identify the single globally best arm $\tilde{I}^* \triangleq \arg\max_{i=1,\cdots,K} \sum_{t=1}^{T} r_{i,t}$. To this end, for each $p_i \in P$, we run a policy that sets $p_t = p_i$ in all rounds and use the method described in Sect. 4.5.4 to compute the payoff $U(t)$ at different checkpoints. By comparing the values of $U(T)$ of different policies, we can determine the best arm. Then for each of the UCB policies, we compute $R(t) \triangleq U^*(t) - U(t)$ at each checkpoint, where $U^*(t)$ is obtained by applying the best-single-arm policy, and $U(t)$ is obtained by applying the UCB policy. The results are shown in Figs. 4.6 and 4.7. When $v_{ori} = 2$, either data

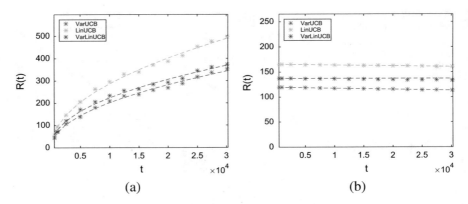

Fig. 4.6 The weak regrets of different learning policies under the setting that data providers' types are uniformly distributed within [0, 1]. The stars represent the actual results, and the dotted lines are fitted curves. (**a**) $v_{ori} = 2$: the fitting function corresponding to the red line is $R(t) = 2.391t^{0.4818}$, the fitting function corresponding to the green line is $R(t) = 3.337t^{0.4838}$, and the fitting function corresponding to the blue line is $R(t) = 3.485t^{0.4522}$; (**b**) $v_{ori} = 10$: the fitting function corresponding to the red line is $R(t) = -0.0002t + 119.5$, the fitting function corresponding to the green line is $R(t) = -0.0002t + 165.5$, and the fitting function corresponding to the blue line is $R(t) = -0.0001t + 137.5$

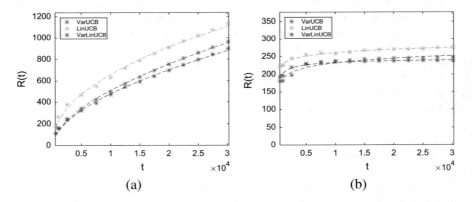

Fig. 4.7 The weak regrets of different learning policies under the setting that data providers' types are approximately normally distributed within [0, 1]. The stars represent the actual results, and the dotted lines are fitted curves. (**a**) $v_{ori} = 2$: the fitting function corresponding to the red line is $R(t) = 2.803t^{0.5652}$, the fitting function corresponding to the green line is $R(t) = 8.751t^{0.4695}$, and the fitting function corresponding to the blue line is $R(t) = 2.984t^{0.5515}$; (**b**) $v_{ori} = 10$: the fitting function corresponding to the red line is $R(t) = 18.46 \ln(28.78t)$, the fitting function corresponding to the green line is $R(t) = 13.15 \ln(41100t)$, and the fitting function corresponding to the blue line is $R(t) = -743t^{-0.3946t} + 252.7$

providers' types are uniformly distributed or normally distributed, the best arm is $p_{\tilde{j}*} = 0.7$. By fitting the simulation results we can see that, given the time horizon T, the payoffs of the proposed policies approach that of the best arm at a rate approximate to $O\left(\sqrt{T}\right)$. When $v_{ori} = 10$ and data providers' types are uniformly distributed, the best arm is $p_{\tilde{j}*} = 1$, and the regrets of the proposed policies almost remains unchanged as time evolves. When $v_{ori} = 10$ and data providers' types are normally distributed, the best arm is $p_{\tilde{j}*} = 0.9$, and for each of the policies, the regret grows at a very low rate. These results imply that when $v_{ori} = 10$, the performance of proposed polices is quite close to that of the best-single-arm policy. Also it should be noted that when $v_{ori} = 10$, the best price $p_{\tilde{j}*}$ is equal or close to the maximum possible value. This coincides with the intuition, since if the value of data is much higher than the maximal price, then it is better for the collector to choose a high price so that most data providers will be encouraged to provide their data.

In practice, the collector may need to announce a single price to all data providers, in case that the data providers feel they are treated unequally. In this situation, the sequential data collecting process investigated in this paper can be seen as the pre-study on privacy valuation of the data providers, based on which the collector can determine the price. The evaluation results of the regrets imply that by applying the proposed learning polices, the collector can find the best price with high probability. In addition to computing the regrets, for each of the three policies, namely VarUCB, LinUCB and VarLinUCB, we check which price is most frequently chosen. And we find that the "best" price picked by the policy coincides with the price that brings the maximal total payoff. This result again demonstrate the effectiveness of the proposed policies.

4.5.5.3 Influence of the Input Parameter

In addition to the comparison of different learning policies, we have conducted simulations to see how the parameter α affects the performance of the UCB-based policies. From the results shown in Figs. 4.8 and 4.9 we can see that, for UCB, LinUCB and VarLinUCB, the overall trend is that the performance gets better as α increases. While the policy VarUCB behaves differently: compared to the other three policies, VarUCB is less sensitive to α; and when $\alpha > 0.3$, the performance declines slowly as α increases.

The different trends shown by UCB and VarUCB can be explained as follows. As descried in Sect. 4.4.1, a large α means it is more likely that the collector takes an explorative action, that is, the collector prefers the price which has not been chosen before rather than the one that has brought most rewards so far. The simulation results demonstrate that instead of directly taking the average of past rewards as the expected reward, the collector can get a more reliable estimate, as we expected, when he uses historical information to estimate the cumulative distribution $F_\theta\left(\cdot\right)$ (see Algorithm 2). Hence, when the policy VarUCB is applied, it is appropriate

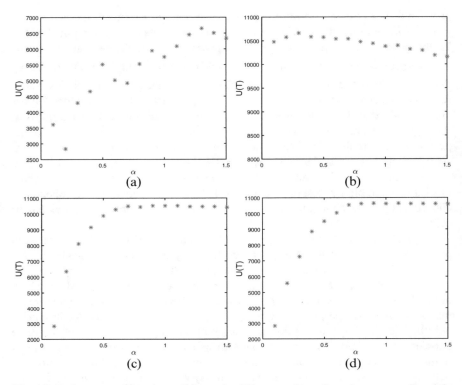

Fig. 4.8 Performance of learning policies under different settings of α, where $v_{ori} = 2$, and the data providers' types are uniformly distributed within $[0, 1]$. (**a**) UCB. (**b**) VarUCB. (**c**) LinUCB. (**d**) VarLinUCB

for the collector to choose a small α to improve his confidence in the knowledge obtained from history (i.e. the estimated expected reward). Besides, from Figs. 4.8a, b, and 4.9a, b we can see that, the performance of UCB is more unstable than that of VarUCB. This result also suggests that VarUCB makes a more reliable estimate of the expected reward.

For LinUCB and VarLinUCB, strictly speaking, the value of α is not properly set in our experiment. According to Theorem1, α is defined as $1 + \sqrt{\frac{1}{2} \ln \frac{2T}{\eta}}$, which means $\alpha > 1$. During the simulation, for comparison purpose, we test the same group of α the for all policies. Though theoretically α should be larger than 1, simulation results show that a small value of α is also feasible. And from Figs. 4.8 and 4.9 we can see that, when α is small, LinUCB and VarLinUCB may perform even worse than the basic policy UCB. While when $\alpha > 1$, LinUCB and VarLinUCB consistently perform much better than UCB, which coincides with our expectation. Another thing should be noted is that after α exceeds 1, both LinUCB and VarLinUCB no longer show significant improvement in performance. This

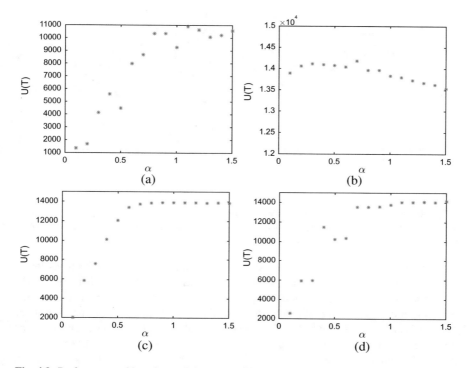

Fig. 4.9 Performance of learning policies under different settings of α, where $v_{ori} = 2$, and the data providers' types are approximately normally distributed within $[0, 1]$. (**a**) UCB. (**b**) VarUCB. (**c**) LinUCB. (**d**) VarLinUCB

result implies that when $\alpha > 1$, the upper confidence bound $\mathbf{x}_t^T \widehat{\boldsymbol{\omega}}_i + \alpha \sqrt{\mathbf{x}_t^T \mathbf{A}_i^{-1} \mathbf{x}_t}$ is accurate enough to show the real differences among different arms' expected rewards, so that the collector can make the best choice at that time, and increasing α will not affect the collector's decision.

4.5.5.4 Influence of the Anonymization Parameter

In previous simulations, we set the parameter of k-anonymity as $k = 10$. To further investigate how the value of k influence the learning results, we conduct another group simulations on the uniformly-distributed data providers. The parameters are set as follows: $k = 50$, $v_{ori} = 2$, and $\alpha \in \{0.1, 0.2, \cdots, 1.5\}$. Similar as before, the best result of each learning policy is reported. As shown in Fig. 4.10a, the three proposed UCB-based polices, namely VarUCB, LinUCB, and VarLinUCB, still show good performance. And by comparing Figs. 4.3a and 4.10a we can see that, for a given policy, the payoff corresponding to $k = 50$ is smaller than that

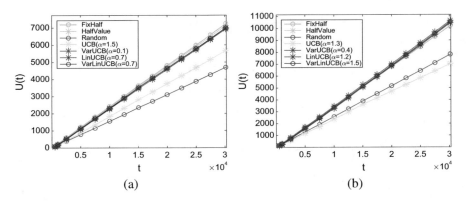

Fig. 4.10 Performance of different learning policies under the setting that data providers' types are uniformly distributed within [0, 1] and $v_{ori} = 2$. (**a**) $k = 50$, $K = 10$. (**b**) $k = 10$, $K = 100$

corresponding to $k = 10$. This is easy to understand, since a larger k causes larger information loss, which means the anonymized data become less valuable to the collector. Also it should be noted that given $k = 50$, the two simple policies, namely FixHalf and HalfValue, perform even better than the proposed UCB-based polices when t is large. As shown in Fig. 4.1b, when $k = 50$ and the size of data is large, the information loss caused by anonymization is about 0.5. Then according to (4.5), the value of v_t approximates to $0.5v_{ori}$ as N_t grows. Therefore, when the data owners' types are uniformly distributed and $v_{ori} = 2$, the price chosen by FixHalf and HalfValue is close to the optimal price $0.5v_t$, especially when t is large.

4.5.5.5 Influence of the Discretization of Price

As described in Sect. 4.5.3, the set of prices is defined as $P \triangleq \{p_i | p_i = \frac{i}{K}, i = 1, \cdots, K\}$ with $K = 10$. In other words, the price range is equally divided into ten subintervals. The discretization of the price simplifies the design of the learning policy, at the cost of sub-optimal prices being found by the policy. Intuitively, if the price range is discretized into more subintervals, the optimal price found by the policy should be closer to the real optimal price. To investigate how the discretization of the price influences the learning result, we conduct a group of simulations on the uniformly-distributed data providers under the setting $k = 10$, $v_{ori} = 2$, and $K = 100$. Simulation results are shown in Fig. 4.10b. By comparing Figs. 4.3a and 4.10b we can see that, given a learning policy, the data collector gets similar payoffs under the two settings. The result that the increase of K does not lead to a significant improvement in the performance of the learning policy implies that instead of equally dividing the price range into more subintervals, we should consider more sophisticated approaches for discretization. We'll investigate this issue in future work.

4.6 Conclusion

Determining the price of personal data is of great importance for implementing the personal data market. In this chapter, we study the pricing problem in a setting where a data collector sequentially interacts with multiple data providers, each of whom has a valuation of privacy that is drawn from an unknown distribution. The pricing problem is formulated as a multi-armed bandit problem. And due to the information loss caused by privacy protection techniques, the distributions of rewards associated to the arms are time-variant. Based on the basic UCB policy, we proposed several learning policies to adapt to the time variant characteristic. To evaluate the performance of the policies, we have conducted simulations under different distributions of data providers' types. Simulation results demonstrate that the proposed learning polices can bring the collector a good payoff. And based on the learning results, the collector can make the best decision if he needs to set a single price for data providers.

Currently, we assume that the collector chooses prices from a discrete set which is obtained by equally dividing the price range. Such a discretization method is easy to implement but may miss the opportunity to find the actual best price. In future work, we will investigate how to adaptively discretize the price range during the learning process, so that the more promising subintervals will be discretized more finely. Besides, considering that the data provider's valuation of privacy may be influenced by the price offered by the collector, which means the data provider may react strategically, we will study how to modify the bandit formulation and the learning policies to deal with such data providers.

Appendix: Proof of Theorem 1

Let $\mathbf{A}_i = \mathbf{D}_i^T \mathbf{D}_i + \mathbf{I}$ and $\alpha = 1 + \sqrt{\frac{1}{2} \ln \frac{2T}{\eta}}$, where η is a positive constant and $\eta < T$. There is

$$
\begin{aligned}
\mathbf{x}_t^T \widehat{\boldsymbol{\omega}}_i - \mathbf{x}_t^T \boldsymbol{\omega}_i^* &= \mathbf{x}_t^T \mathbf{A}_i^{-1} \mathbf{D}_i^T \mathbf{c}_i - \mathbf{x}_t^T \boldsymbol{\omega}_i^* \\
&= \mathbf{x}_t^T \mathbf{A}_i^{-1} \mathbf{D}_i^T \mathbf{c}_i - \mathbf{x}_t^T \mathbf{A}_i^{-1} \mathbf{A}_i \boldsymbol{\omega}_i^* \\
&= \mathbf{x}_t^T \mathbf{A}_i^{-1} \mathbf{D}_i^T \mathbf{c}_i - \mathbf{x}_t^T \mathbf{A}_i^{-1} \left(\mathbf{D}_i^T \mathbf{D}_i + \mathbf{I} \right) \boldsymbol{\omega}_i^* \\
&= \mathbf{x}_t^T \mathbf{A}_i^{-1} \mathbf{D}_i^T \left(\mathbf{c}_i - \mathbf{D}_i \boldsymbol{\omega}_i^* \right) - \mathbf{x}_t^T \mathbf{A}_i^{-1} \boldsymbol{\omega}_i^*.
\end{aligned}
\tag{4.20}
$$

Then we get

$$
\left| \mathbf{x}_t^T \widehat{\boldsymbol{\omega}}_i - \mathbf{x}_t^T \boldsymbol{\omega}_i^* \right| \leq \left| \mathbf{x}_t^T \mathbf{A}_i^{-1} \mathbf{D}_i^T \left(\mathbf{c}_i - \mathbf{D}_i \boldsymbol{\omega}_i^* \right) \right| + \left| \mathbf{x}_t^T \mathbf{A}_i^{-1} \boldsymbol{\omega}_i^* \right|.
\tag{4.21}
$$

For the first part in the right-hand side of above inequality: notice that $\mathbf{D}_i \boldsymbol{\omega}_i^* = E[\mathbf{c}_i]$, by applying McDiarmid's inequality [33] we get

$$\Pr\left(\left|\mathbf{x}_t^T \mathbf{A}_i^{-1} \mathbf{D}_i^T \left(\mathbf{c}_i - \mathbf{D}_i \boldsymbol{\omega}_i^*\right)\right| > \varepsilon \sqrt{\mathbf{x}_t^T \mathbf{A}_i^{-1} \mathbf{x}_t}\right)$$

$$\leq 2 \exp\left(-\frac{2\varepsilon^2 \mathbf{x}_t^T \mathbf{A}_i^{-1} \mathbf{x}_t}{\left\|\mathbf{x}_t^T \mathbf{A}_i^{-1} \mathbf{D}_i^T\right\|^2}\right)$$

$$= 2 \exp\left(-\frac{2\varepsilon^2 \mathbf{x}_t^T \mathbf{A}_i^{-1} \left(\mathbf{I} + \mathbf{D}_i^T \mathbf{D}_i\right) \mathbf{A}_i^{-1} \mathbf{x}_t}{\left\|\mathbf{x}_t^T \mathbf{A}_i^{-1} \mathbf{D}_i^T\right\|^2}\right)$$

$$\leq 2 \exp\left(-\frac{2\varepsilon^2 \mathbf{x}_t^T \mathbf{A}_i^{-1} \mathbf{D}_i^T \mathbf{D}_i \mathbf{A}_i^{-1} \mathbf{x}_t}{\left\|\mathbf{x}_t^T \mathbf{A}_i^{-1} \mathbf{D}_i^T\right\|^2}\right) \qquad (4.22)$$

$$= 2 \exp\left(-\frac{2\varepsilon^2 \left\|\mathbf{x}_t^T \mathbf{A}_i^{-1} \mathbf{D}_i^T\right\|^2}{\left\|\mathbf{x}_t^T \mathbf{A}_i^{-1} \mathbf{D}_i^T\right\|^2}\right)$$

$$= 2 \exp\left(-2\varepsilon^2\right).$$

For the second part in the right-hand side of (4.21): since $\left\|\boldsymbol{\omega}_i^*\right\| \leq 1$,

$$\left|\mathbf{x}_t^T \mathbf{A}_i^{-1} \boldsymbol{\omega}_i^*\right| \leq \left\|\mathbf{x}_t^T \mathbf{A}_i^{-1}\right\| \left\|\boldsymbol{\omega}_i^*\right\|$$

$$\leq \left\|\mathbf{x}_t^T \mathbf{A}_i^{-1}\right\|$$

$$= \sqrt{\mathbf{x}_t^T \mathbf{A}_i^{-1} \mathbf{A}_i^{-1} \mathbf{x}_t} \qquad (4.23)$$

$$\leq \sqrt{\mathbf{x}_t^T \mathbf{A}_i^{-1} \left(\mathbf{D}_i^T \mathbf{D}_i + \mathbf{I}\right) \mathbf{A}_i^{-1} \mathbf{x}_t}$$

$$= \sqrt{\mathbf{x}_t^T \mathbf{A}_i^{-1} \mathbf{x}_t}.$$

Let $\Delta_i = \sqrt{\mathbf{x}_t^T \mathbf{A}_i^{-1} \mathbf{x}_t}$. By using (4.22) and (4.23) we get

$$\Pr\left(\left|\mathbf{x}_t^T \widehat{\boldsymbol{\omega}}_i - \mathbf{x}_t^T \boldsymbol{\omega}_i^*\right| > (\varepsilon + 1)\,\Delta_i\right)$$

$$\leq \Pr\left(\left|\mathbf{x}_t^T \mathbf{A}_i^{-1}\mathbf{D}_i^T\left(\mathbf{c}_i - \mathbf{D}_i\boldsymbol{\omega}_i^*\right)\right| + \left|\mathbf{x}_t^T \mathbf{A}_i^{-1}\boldsymbol{\omega}_i^*\right| > (\varepsilon + 1)\,\Delta_i\right) \quad (4.24)$$

$$\leq 2\exp\left(-2\varepsilon^2\right).$$

References

1. L. Xu, C. Jiang, J. Wang, J. Yuan, and Y. Ren, "Information security in big data: Privacy and data mining," *IEEE Access*, vol. 2, pp. 1149–1176, 2014.
2. L. Xu, C. Jiang, Y. Chen, J. Wang, and Y. Ren, "A framework for categorizing and applying privacy-preservation techniques in big data mining," *Computer*, vol. 49, no. 2, pp. 54–62, Feb 2016.
3. B. Fung, K. Wang, R. Chen, and P. S. Yu, "Privacy-preserving data publishing: A survey of recent developments," *ACM Computing Surveys (CSUR)*, vol. 42, no. 4, p. 14, 2010.
4. R. Agrawal and R. Srikant, "Privacy-preserving data mining," *SIGMOD Rec.*, vol. 29, no. 2, pp. 439–450, 2000.
5. A. Acquisti, C. R. Taylor, and L. Wagman, "The economics of privacy," http://ssrn.com/abstract=2580411, March 2015.
6. A. Roosendaal, M. van Lieshout, and A. F. van Veenstra, "Personal data markets," TNO, Delft, report TNO 2014 R11390, November 2014.
7. L. Xu, C. Jiang, Y. Chen, Y. Ren, and K. Liu, "Privacy or utility in data collection? a contract theoretic approach," *Selected Topics in Signal Processing, IEEE Journal of*, vol. 9, no. 7, pp. 1256–1269, Oct 2015.
8. L. Xu, C. Jiang, Y. Qian, Y. Zhao, J. Li, and Y. Ren, "Dynamic privacy pricing: A multi-armed bandit approach with time-variant rewards," *IEEE Transactions on Information Forensics and Security*, vol. 12, no. 2, pp. 271–285, Feb 2017.
9. A. Singla and A. Krause, "Truthful incentives in crowdsourcing tasks using regret minimization mechanisms," in *Proceedings of the 22Nd International Conference on World Wide Web*, ser. WWW '13. New York, NY, USA: ACM, 2013, pp. 1167–1178.
10. K. Amin, A. Rostamizadeh, and U. Syed, "Learning prices for repeated auctions with strategic buyers," in *Advances in Neural Information Processing Systems 26*, C. J. C. Burges, L. Bottou, M. Welling, Z. Ghahramani, and K. Q. Weinberger, Eds. Curran Associates, Inc., 2013, pp. 1169–1177. [Online]. Available: http://papers.nips.cc/paper/5010-learning-prices-for-repeated-auctions-with-strategic-buyers.pdf
11. S. Bubeck and N. Cesa-Bianchi, "Regret analysis of stochastic and nonstochastic multi-armed bandit problems," *arXiv preprint arXiv:1204.5721*, 2012.
12. P. Auer, N. Cesa-Bianchi, Y. Freund, and R. E. Schapire, "The nonstochastic multiarmed bandit problem," *SIAM Journal on Computing*, vol. 32, no. 1, pp. 48–77, 2002.
13. C. Li, D. Y. Li, G. Miklau, and D. Suciu, "A theory of pricing private data," *ACM Trans. Database Syst.*, vol. 39, no. 4, pp. 34:1–34:28, Dec. 2014.
14. P. Koutris, P. Upadhyaya, M. Balazinska, B. Howe, and D. Suciu, "Query-based data pricing," *J. ACM*, vol. 62, no. 5, pp. 43:1–43:44, Nov. 2015.
15. B.-R. Lin and D. Kifer, "On arbitrage-free pricing for general data queries," *Proc. VLDB Endow.*, vol. 7, no. 9, pp. 757–768, May 2014.
16. V. Gkatzelis, C. Aperjis, and B. A. Huberman, "Pricing private data," *Electronic Markets*, vol. 25, no. 2, pp. 109–123, 2015.
17. X.-B. Li and S. Raghunathan, "Pricing and disseminating customer data with privacy awareness," *Decision Support Systems*, vol. 59, pp. 63–73, March 2014.

18. A. V. den Boer, "Dynamic pricing and learning: historical origins, current research, and new directions," *Surveys in operations research and management science*, vol. 20, no. 1, pp. 1–18, 2015.
19. L. Xiao, Y. Li, J. Liu, and Y. Zhao, "Power control with reinforcement learning in cooperative cognitive radio networks against jamming," *The Journal of Supercomputing*, vol. 71, no. 9, pp. 3237–3257, 2015.
20. J. Liu, L. Xiao, G. Liu, and Y. Zhao, "Active authentication with reinforcement learning based on ambient radio signals," *Multimedia Tools and Applications*, pp. 1–20, 2015.
21. L. Xiao, Y. Li, G. Han, G. Liu, and W. Zhuang, "Phy-layer spoofing detection with reinforcement learning in wireless networks," *IEEE Transactions on Vehicular Technology*, vol. PP, no. 99, pp. 1–1, 2016.
22. O. Besbes, Y. Gur, and A. Zeevi, "Stochastic multi-armed-bandit problem with non-stationary rewards," in *Advances in Neural Information Processing Systems*, 2014, pp. 199–207.
23. S. Vakili, Q. Zhao, and Y. Zhou, "Time-varying stochastic multi-armed bandit problems," in *Signals, Systems and Computers, 2014 48th Asilomar Conference on*. IEEE, 2014, pp. 2103–2107.
24. A. Garivier and E. Moulines, "On upper-confidence bound policies for switching bandit problems," in *Algorithmic Learning Theory*. Springer, 2011, pp. 174–188.
25. L. SWEENEY, "Achieving k-anonymity privacy protection using generalization and suppression," *International Journal of Uncertainty, Fuzziness and Knowledge-Based Systems*, vol. 10, no. 05, pp. 571–588, 2002.
26. P. Auer, "Using confidence bounds for exploitation-exploration trade-offs," *The Journal of Machine Learning Research*, vol. 3, pp. 397–422, 2003.
27. P. Auer, N. Cesa-Bianchi, and P. Fischer, "Finite-time analysis of the multiarmed bandit problem," *Machine learning*, vol. 47, no. 2–3, pp. 235–256, 2002.
28. L. Li, W. Chu, J. Langford, and R. E. Schapire, "A contextual-bandit approach to personalized news article recommendation," in *Proceedings of the 19th international conference on World wide web*. ACM, 2010, pp. 661–670.
29. K. Bache and M. Lichman, "UCI machine learning repository," 2013. [Online]. Available: http://archive.ics.uci.edu/ml
30. K. LeFevre, D. J. DeWitt, and R. Ramakrishnan, "Incognito: Efficient full-domain k-anonymity," in *Proceedings of the 2005 ACM SIGMOD International Conference on Management of Data*, ser. SIGMOD '05. New York, NY, USA: ACM, 2005, pp. 49–60. [Online]. Available: http://doi.acm.org/10.1145/1066157.1066164
31. F. Kohlmayer, F. Prasser, C. Eckert, A. Kemper, and K. Kuhn, "Flash: Efficient, stable and optimal k-anonymity," in *Privacy, Security, Risk and Trust (PASSAT), 2012 International Conference on and 2012 International Confernece on Social Computing (SocialCom)*, Sept 2012, pp. 708–717.
32. V. S. Iyengar, "Transforming data to satisfy privacy constraints," in *Proceedings of the Eighth ACM SIGKDD International Conference on Knowledge Discovery and Data Mining*, ser. KDD '02. New York, NY, USA: ACM, 2002, pp. 279–288. [Online]. Available: http://doi.acm.org/10.1145/775047.775089
33. C. McDiarmid, "Concentration," in *Probabilistic methods for algorithmic discrete mathematics*. Springer, 1998, pp. 195–248.

Chapter 5
User Participation Game in Collaborative Filtering

Abstract One of the most important applications of data mining is personalized recommendation. User participation plays a vital role in personalized recommendation systems, especially those based on collaborative filtering techniques. A user can get high-quality recommendations only when both the user himself/herself and other users actively participate, i.e. providing sufficient rating data. However, due to the rating cost, e.g. the privacy loss, rational users tend to provide as few ratings as possible. There is a trade-off between the rating cost and the recommendation quality. In this chapter, we model the interactions among users as a game in satisfaction form and study the corresponding equilibrium, namely satisfaction equilibrium (SE). Considering that accumulated ratings are used for generating recommendations, we design a behavior rule which allows users to achieve an SE via iteratively rating items. We theoretically analyze under what conditions an SE can be learned via the behavior rule. Experimental results demonstrate that, if all users have moderate expectations for recommendation quality and satisfied users are willing to provide more ratings, then all users can get satisfying recommendations without providing many ratings.

5.1 Introduction

5.1.1 Collaborative Filtering-Based Recommendation

Recommendation system has been successfully applied in a variety of applications [1]. The predominant approach to building recommendation systems is collaborative filtering (CF) [2], where the key idea is to utilize the *ratings* collected from users to identify users with similar interests and to predict which items the users may be interested in. Conventionally, ratings are organized into a user-item matrix $\mathbf{R} = [r_{ij}]_{N \times M}$ with the rating r_{ij} indicating user i's preference for item j. The task of the recommendation server (RS) is to predict the missing values in the matrix.

Users' rating data are the fundamental resources of CF-based recommendation systems, which means user participation is of vital importance for the success of

© Springer International Publishing AG, part of Springer Nature 2018

L. Xu et al., *Data Privacy Games*, https://doi.org/10.1007/978-3-319-77965-2_5

recommendation. Generally, a user assigns ratings to items after he[1] has obtained experience of the items. In practice, the number of total items available for recommendation is much larger than the number of items that a user has experienced, thus the rating matrix is sparse. To make things worse, due to the cost incurred by rating items, such as time consumption and privacy disclosure, users will not rate every item that they have experienced. The insufficiency of rating data inevitably impairs the recommendation quality [3].

5.1.2 Encourage User Participation

To deal with the aforementioned problem, researchers have proposed various approaches, such as exploring the content information [4] and user relationships [5]. Apart from improving the recommendation algorithms [6–8], one can circumvent the problem by designing incentive mechanisms to encourage user participation. Though mechanisms proposed particularly for recommendation systems are rare, incentive mechanisms have been extensively studied in similar contexts such as peer-to-peer resource sharing [9], crowdsourcing [10], cooperation in wireless communications [11], etc.

In recommendation systems, the RS can offer various incentives to users so as to compensate their rating cost. In addition to monetary rewards and other forms of external incentives, the recommendation quality can be considered as an intrinsic incentive for users to rate items. In this chapter we'd like to investigate the influence of the recommendations themselves on users' rating behaviors. Specifically, we are interested in the following questions: whether users, motivated by recommendation quality solely, can contribute sufficient rating data so that the RS can generate *satisfying* recommendations for all users? How should users behave so that the cost of rating and the quality of recommendations can be balanced?

5.1.3 Game-Theoretic Approach

Intuitively, a user may get better recommendations if he reveals more information about his/her preferences to the RS by rating more items, while in the meantime, the user has to pay more cost. When deciding whether to rate an item or not, a user needs to make a trade-off between the cost of rating and the quality of recommendation. Moreover, as the name *collaborative filtering* suggests, whether a user can get good recommendations depends not only on the ratings provided by the user himself, but also on the ratings provided by others. Therefore, interactions of individuals' rating behaviors should be considered when one makes decisions

---

[1]For ease of description, in this chapter we sometimes use *he* to refer to the user.

on rating. Furthermore, users are usually rational, in the sense that a user wishes to obtain good recommendations without rating many items. In such a case, it is natural to employ game theory [12] to model the interactions among users in a CF system.

In this chapter, we build a game theoretical model to study users' rating behaviors in a CF-based recommendation system [13]. Application of game theory has been seen in a few studies of user behavior in a context where individuals' behaviors affect each other [14–16]. Particularly, Halkidi et al. [14] employed game theory to model the interactions among users in a recommendation system. They developed a mathematical framework to address the trade-off between privacy preservation and high-quality recommendation. Different from their study, we model the interactions among users as a satisfactory game with incomplete information: each user only has the knowledge of his own ratings and recommendations, while others' ratings cannot be observed. Meanwhile, the CF algorithm adopted by the RS is also unknown to users. Inspired by Perlaza et al.'s work [17], we apply the notion of *satisfaction equilibrium* (SE), which was originally introduced by Ross and Chaib-draa [18], to analyze the game with incomplete information. A game is said to be in SE when all players simultaneously satisfy their individual constrains. In the context of CF, a user's expectation for recommendation quality is seen as his constrain.

The proposed game is a game with incomplete information. Hence, different from the equilibrium concepts in the context of complete information games, the satisfaction equilibrium arises as a result of a learning process, rather than the result of rational thinking on players' beliefs and observations [18]. Based on the characteristics of recommendation systems, we design a learning algorithm which allows users to achieve an SE. Convergence of the proposed learning algorithm is analyzed theoretically. And we conduct a series of experiments on the Jester data set and the MovieLens data set to verify the feasibility of the learning algorithm. We think that the derived convergence conditions can provide some implications to the design of external incentives.

The rest of the chapter is organized as follows. Section 5.2 presents the experimental proof for the basic assumption based on which we build the game model. Section 5.3 briefly describes the system model while Sect. 5.4 presents in details the game formulation. In Sect. 5.5, we introduce the proposed learning algorithm to achieve the satisfaction equilibrium. The convergence analysis is conducted in Sect. 5.6. Finally, simulation results are shown in Sect. 5.7 and conclusions are drawn in Sect. 5.8.

5.2 Preliminary Analysis

A fundamental assumption of our study is that given the items that the users have experienced and the recommendation algorithm adopted by the RS, the quality of recommendations increases as users provide more ratings. This assumption is quite general. Yet in order to make the chapter more rigorous, we have conducted some

Fig. 5.1 Illustration of the experiment result $\mathbf{Q} = \left[q_{ij}\right]$. Each row of $\mathbf{Q}$ corresponds to a given value of σ_R ($\sigma_R = \frac{70}{100} \cdot \frac{k}{100}, k = 1, 2, \cdots, 100$). Each column of $\mathbf{Q}$ corresponds to a user. The element q_{ij} represents the corresponding recommendation quality. Different values of q_{ij} are indicated by different colors: blue represents low quality, red represents high quality

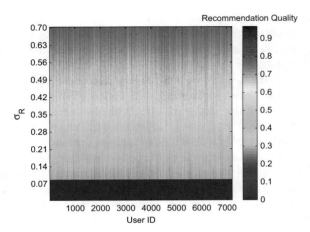

simple experiments to verify the assumption. Experiments were performed on a set of ratings chosen from the Jester data set [19]. Given the original rating matrix, we randomly set some non-zero elements to "0" (denoting missing values). Let σ_R denote the ratio of remaining non-zero elements to original non-zero elements. By this way, we can observe how the recommendation quality changes with the number of ratings. Detailed information about the rating data will be presented in Sect. 5.7.

Experiment results are stored in a matrix $\mathbf{Q} = \left[q_{ij}\right]$, where each column represents a user, each row represents a particular value of σ_R, and q_{ij} denotes the corresponding recommendation quality. Figure 5.1 shows the matrix $\mathbf{Q}$ obtained by applying a user-based CF algorithm, and the recommendation quality is measured by the difference between the predicated ratings and the user's true preferences. Details of the recommendation algorithm and the evaluation metric of recommendation quality will be presented in Sect. 5.3. As we can see, as more ratings are available (σ_R increases), the recommendation quality improves.

To better illustrate the change of recommendation quality, for each value of σ_R we compute the average of the recommendation quality perceived by all users. Figure 5.2 shows the results obtained under different settings of recommendation algorithms and evaluation metrics. It is clear that for any given recommendation algorithm, the recommendation quality improves with the increase of σ_R. Suppose that each user has an expectation for the recommendation quality, then from Fig. 5.2 we can learn that, if users don't have high expectations, a relatively small number of ratings (e.g. $\sigma_R = 0.5$) will be enough to generate satisfying recommendations. With these preliminary results, we can proceed to formal study of the user participation problem.

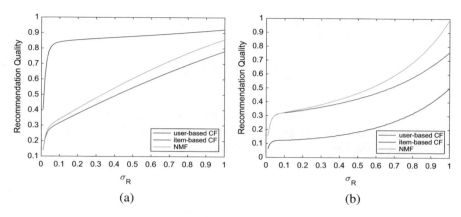

Fig. 5.2 The recommendation quality changes with the number of ratings. Each curve corresponds to one of the following three recommendation algorithms: user-based collaborative filtering [20], item-based collaborative filtering [21], and non-negative matrix factorization [22]. (**a**) The recommendation quality is evaluated by the difference between the predicated ratings and the user's true preferences (see (5.4)). (**b**) The recommendation quality is evaluated by the overlap between the recommended items and the items that the user is mostly interested in (see (5.3))

5.3   System Model

Consider a CF system where a set of users $\mathcal{N} = \{1, 2, \cdots, N\}$ interact with a RS. The RS maintains information about a set of items $S = \{s_1, s_2, \cdots, s_M\}$. Each user experiences a set of items and assigns ratings to some of them. Let S_i and $\tilde{S}_i$ denote the set of items that user i has experienced and rated, respectively, then we have $\tilde{S}_i \subseteq S_i \subseteq S$. From the perspective of the RS, a rating vector $\mathbf{r}_i = (r_{i1}, r_{i2}, \cdots, r_{iM})$ is provided by user i when a set $\tilde{S}_i$ is chosen. We define $r_{ij} \in (0, r_{\max}]$ if $s_j \in \tilde{S}_i$, $r_{ij} = 0$ if $s_j \notin \tilde{S}_i$ $(j = 1, \cdots, M)$. Usually, a high value of r_{ij} implies user i has a strong preference for item s_j.

As mentioned in Sect. 5.1, the reason that the user will not rate all the items in S_i is the time consumption and the privacy loss incurred by rating. In order to protect privacy, the user can provide fake ratings to the RS [23, 24], so that the true preferences of the user will not be disclosed. However, considering that the recommendation quality will be hurt by fake ratings, falsifying ratings can be nontrivial and time-consuming. Also, in practical recommendation systems, the profile of user interest is often represented by some kind of distribution over different types of items [25], which means the values of ratings have little influence on user profile. What matters more is whether the user has rated an item. Hence in this chapter, we assume that as long as the user decides to rate an item, the user will provide a rating that coincides with his true preference.

The ratings provided by all users form a rating matrix $\mathbf{R} = \left[r_{ij} \right]_{N \times M}$. The RS applies some recommendation algorithm to $\mathbf{R}$ to predict users' preferences for those unrated items. A recommendation vector $\hat{\mathbf{r}}_i = (\hat{r}_{i1}, \cdots, \hat{r}_{iM})$ is computed for each user i, where $\hat{r}_{ij}$ is defined as follows:

$$\hat{r}_{ij} = \begin{cases} r_{ij}, & if \ r_{ij} \neq 0 \\ f_{ij}(\mathbf{R}), & if \ r_{ij} = 0 \end{cases}, \tag{5.1}$$

with $f_{ij}(\mathbf{R})$ being the predicted rating determined by both the recommendation algorithm and the whole ratings. For example, if user-oriented neighborhood-based CF [20] is applied, then $f_{ij}(\mathbf{R})$ can be defined as:

$$f_{ij}(\mathbf{R}) = \frac{\sum\limits_{k \in Neighbour(i)} r_{kj} F_{sim}(i, k)}{\sum\limits_{k \in Neighbour(i)} F_{sim}(i, k)}, \tag{5.2}$$

where $F_{sim}(i, k)$ represents the similarity between user i and user k, $Neighbour(i)$ represents the set of users who are most similar to user i. The similarity $F_{sim}(i, k)$ can be measured by Pearson correlation or vector cosine similarity [2].

After computing the recommendation vector, generally the RS will select several items with high $f_{ij}(\mathbf{R})$ and recommend them to the user. Then the user can evaluate whether the recommended items match his interest. Let $\mathbf{p}_i = (p_{i1}, \cdots, p_{iM})$ denote user i's interest, where p_{ij} represents user i's true preference for item s_j ($j = 1, \cdots, M$). We assume $0 \leq p_{ij} \leq r_{\max}$ and define $p_{ij} = r_{ij}$ for $s_j \in \tilde{S}_i$. Let $\hat{S}_i$ denote the set of K items recommended by the RS. Let $\breve{S}_i$ denote the set of K items that correspond to the K highest p_{ij} in the set $S \backslash S_i$. That is, $\breve{S}_i$ denote the set of items that user i has not experienced yet but is interest in. Then the quality of the recommendation result $\hat{S}_i$, denoted as $QoR\left(\hat{S}_i\right)$, can be defined as

$$QoR\left(\hat{S}_i\right) = \frac{\left|\hat{S}_i \cap \breve{S}_i\right|}{K}, \tag{5.3}$$

where $|A|$ denote the cardinality of the set A.

In the study of recommendation systems, the recommendation quality is often evaluated by mean absolute error (MAE) or root mean squared error (RMSE) [2]. Based on the definition of RMSE, we assume that the RS returns the whole vector $\hat{\mathbf{r}}_i$ to the user, and the quality of $\hat{\mathbf{r}}_i$ is evaluated by a user-specific function $g_i : \mathbb{R}^M \to \mathbb{R}$ which is defined as

$$g_i\left(\hat{\mathbf{r}}_i\right) = 1 - \frac{\sqrt{\sum\limits_{j=1}^{M} \left(\hat{r}_{ij} - p_{ij}\right)^2}}{r_{\max}\sqrt{M}}. \tag{5.4}$$

A large $g_i\left(\hat{\mathbf{r}}_i\right)$ implies high similarity between $\mathbf{r}_i$ and $\mathbf{p}_i$, namely high recommendation quality. In subsequent analysis, we mainly use (5.4) as the metric of recommendation quality. From (5.1), (5.3) and (5.4), we can see that the recommendation quality obtained by one user is affected by other users' ratings. In other words, users in a CF system interact with each other via providing ratings to the RS. In the following section, we will use satisfactory game to formulate the interaction among users.

5.4 Satisfactory Game Formulation

5.4.1 Players and Actions

We consider all the users in $\mathcal{N}$ as players and the set $\tilde{S}_i$ as user i's action, i.e., $a_i = \tilde{S}_i$. Let $\mathcal{A}_i$ denote the action space of user i. All users share the same action space, i.e. for any $i \in \mathcal{N}$, there is $\mathcal{A}_i = \left(A^{(1)}, \cdots, A^{(K)}\right)$, where $K = 2^{|S|} - 1$, $A^{(k)} \subseteq S$ ($k = 1, \cdots, K$) and $A^{(k)} \neq \varnothing$. When choosing an action, each user follows his own probability distribution over the action space. We use $\boldsymbol{\pi}_i = \left(\pi_i^{(1)}, \cdots, \pi_i^{(K)}\right)$ to denote the distribution, where $\pi_i^{(k)} \triangleq \Pr\left(a_i = A^{(k)}\right)$ represents the probability that user i chooses the action $A^{(k)}$.

Given an action profile $\mathbf{a} = (a_1, \cdots, a_N) \in \mathcal{A}$ ($\mathcal{A} = \mathcal{A}_1 \times \cdots \times \mathcal{A}_N$), the rating matrix $\mathbf{R}$ obtained by the RS is determined. Considering that the recommendation $\hat{\mathbf{r}}_i$ is fully determined by $\mathbf{R}$ when the recommendation algorithm is specified, we introduce a mapping $h_i : \mathcal{A} \to \mathbb{R}$ to show the influence of users' actions on recommendation quality:

$$g_i\left(\hat{\mathbf{r}}_i\right) = h_i\left(\mathbf{a}\right) = h_i\left(a_i, \mathbf{a}_{-i}\right) , \tag{5.5}$$

where $\mathbf{a}_{-i} = (a_1, \cdots, a_{i-1}, a_{i+1}, \cdots, a_N) \in \mathcal{A}_{-i}$, $\mathcal{A}_{-i} = \mathcal{A}_1 \times \cdots \mathcal{A}_{i-1} \times \mathcal{A}_{i+1} \cdots \times \mathcal{A}_N$.

Intuitively, either the user i himself or other users rate more items, the rating matrix will become less sparse, and user i can get better recommendations. We introduce the notion of *rating completeness* to measure the relative amount of ratings provided by the user. Given the set S_i, user i's rating completeness σ_i is defined as

$$\sigma_i = \frac{|a_i|}{|S_i|} . \tag{5.6}$$

Notice that $a_i \subseteq S_i$ and $a_i \neq \varnothing$, hence $0 < \sigma_i \leq 1$. A large σ_i means user i actively participates in the rating activity. We use σ_{-i} to denote the average of other users' rating completeness:

$$\sigma_{-i} = \frac{1}{N-1} \sum_{j \in \mathcal{N}, \, j \neq i} \sigma_j . \tag{5.7}$$

By introducing σ_i and σ_{-i}, we can rewrite $h_i\,(a_i, \mathbf{a}_{-i})$ as

$$h_i\,(a_i, \mathbf{a}_{-i}) = h\,(\sigma_i, \sigma_{-i}; \mathbf{p}_i)\,, \tag{5.8}$$

where the function $h\,(\cdot; \mathbf{p}_i)$ with parameter $\mathbf{p}_i$ takes σ_i and σ_{-i} as input.

As mentioned in Sect. 5.1, rating items incurs some cost. The more items the user rates, the higher cost he has to pay. Let $c_i\,(a_i)$ denote the cost paid by user i when he chooses the action a_i, then for any $a'_i \in \mathscr{A}_i$, $a''_i \in \mathscr{A}_i$, if $a'_i \subset a''_i$, there is $c_i\left(a'_i\right) < c_i\left(a''_i\right)$.

5.4.2   Satisfaction Form

Due to the rating cost, usually the user will not rate all the items he has experienced. As we have discussed in Sect. 5.2, given the recommendation algorithm, the evaluation metric of recommendation quality, and the items that users have experienced, the recommendation quality perceived by every user increases with the number of ratings provided by users. This means that when every user has rated all the items he has experienced, i.e. each user i chooses the action $a^*_i \triangleq S_i$, every user can receive the best recommendation that he can get. In such a case, the rating completeness of every user is 1. If we use $\Gamma_i^{\max}$ to denote the best recommendation quality, then there is

$$\Gamma_i^{\max} = h\,(1, 1; \mathbf{p}_i)\,. \tag{5.9}$$

In most cases, the rating completeness of a user is less than 1, hence the best result $\Gamma_i^{\max}$ can hardly be realized. Suppose that each user i has a relatively low expectation Γ_i ($\Gamma_i < \Gamma_i^{\max}$) for the recommendation quality. Given an action profile $\mathbf{a}$, as long as $h_i\,(\mathbf{a}) \geq \Gamma_i$, user i will be *satisfied*.

From (5.5) we know that, given the actions of other users, certain actions should be chosen by user i so that user i can get satisfying recommendations. We use $f_i\,(\mathbf{a}_{-i})$ to denote the set of such actions:

$$f_i\,(\mathbf{a}_{-i}) = \{a_i \in \mathscr{A}_i : h_i\,(a_i, \mathbf{a}_{-i}) \geq \Gamma_i\}\,. \tag{5.10}$$

For any $\mathbf{a}_{-i} \in \mathscr{A}_{-i}$, the mapping $f_i : \mathscr{A}_{-i} \to 2^{\mathscr{A}_i}$ determines the actions available for user i to satisfy his expectation. It should be noted that, for some $\mathbf{a}_{-i}$, $f_i\,(\mathbf{a}_{-i})$ may be empty. For example, suppose that each user in $\mathscr{N}$, expect user i, rates only one item. Then even if user i rates all the items he has experienced, the ratings are not enough to reflect the real similarities between users. Consequently, user i cannot get satisfying recommendations.

Based on above discussions, we can describe the proposed game by the following triplet:

$$\hat{G}_{CF} = \left(\mathcal{N}, \{ \mathscr{A}_i \}_{i \in \mathcal{N}}, \{ f_i \}_{i \in \mathcal{N}} \right) \tag{5.11}$$

This formulation of game is called *satisfaction form*, which was first introduced by Perlaza et al. [17] to model the problem of quality-of-service provisioning in decentralized self-configuring networks.

5.4.3 Satisfaction Equilibrium

An important outcome of a game in satisfaction form is the one where all players are satisfied. This outcome is referred to as *satisfaction equilibrium* (SE) [17]:

Definition 1 (Satisfaction Equilibrium) An action profile $\mathbf{a}^+$ is an equilibrium for the game $\hat{G}_{CF} = \left(\mathcal{N}, \{ \mathscr{A}_i \}_{i \in \mathcal{N}}, \{ f_i \}_{i \in \mathcal{N}} \right)$, if $\forall i \in \mathcal{N}$, there is $a_i^+ \in f_i \left(\mathbf{a}_{-i}^+ \right)$.

We have assumed that for all $i \in \mathcal{N}$, there is $\Gamma_i < \Gamma_i^{\max}$, hence the action profile $\mathbf{a}^* \triangleq (S_1, S_2, \cdots, S_N)$ is an SE of the proposed game. However, $\mathbf{a}^*$ requires every user to pay the highest cost $c_i (S_i)$, which may exceed the necessary cost for achieving user's expectation. It is more practical to find a lower-cost SE $\mathbf{a}^+ = \left(a_1^+, \cdots, a_N^+ \right)$ which satisfies:

(i) $\forall i \in \mathcal{N}$, there is $a_i^+ \in f_i \left(\mathbf{a}_{-i}^+ \right)$ and $c_i \left(a_i^+ \right) \leq c_i (S_i)$;
(ii) there is at least one user who doesn't have to provide his complete ratings, that is, $\exists i \in \mathcal{N}$, $c_i \left(a_i^+ \right) < c_i (S_i)$.

5.5 Learning Satisfaction Equilibrium

The game described above is a game with incomplete information, since each user has no knowledge of other users' actions. Different from general equilibrium concepts of games with complete information, the satisfaction equilibrium is obtained as the result of a learning process, rather than the result of rational thinking on players' beliefs and observations [18]. In this section, we study the behavior rule that allows users to learn a satisfaction equilibrium. The equilibrium learning is essentially an iterative process of information exchange between users and the RS. For the RS, the iterative process provides a way to acquire a certain amount of information to build a profile for a user [26, 27]. During the learning process, each user chooses his actions as follows.

Initially, user i chooses an action $a_i(0)$ based on the probability distribution $\pi_i(0) = \left(\pi_i^{(1)}(0), \cdots, \pi_i^{(K)}(0)\right)$, where for any $k \in \{1, \cdots, K\}$, $\pi_i^{(k)}(0)$ is defined as follows:

$$\pi_i^{(k)}(0) = \begin{cases} \beta_i(0)/\alpha^{c_i\left(A^{(k)}\right)}, & if \quad A^{(k)} \subseteq S_i \\ 0, & otherwise \end{cases}. \tag{5.12}$$

where parameter $\alpha > 1$ shows how much users care about the cost. A large α means it is more likely that the user will rate a small number of new items (i.e. unrated items) in one iteration. On the other hand, if α is small, users may provide sufficient ratings in a few iterations, which means an SE can be quickly achieved. The normalization factor $\beta_i(0)$ is defined as:

$$\beta_i(0) = \frac{1}{\displaystyle\sum_{k:\ A^{(k)} \subseteq S_i} \alpha^{-c_i\left(A^{(k)}\right)}}. \tag{5.13}$$

After every user has chosen his action, the RS computes the recommendations based on the initial rating matrix $\mathbf{R}(0)$ and returns $\hat{\mathbf{r}}_i(0)$ to user i.

At the beginning of iteration n ($n = 1, 2, \cdots$), user i evaluates $\hat{\mathbf{r}}_i(n-1)$ to see whether it is satisfactory. We use a binary variable $v_i(n-1)$ to indicate the evaluation result:

$$v_i(n-1) = \begin{cases} 1, & if \quad g_i\left(\hat{\mathbf{r}}_i(n-1)\right) \geq \Gamma_i \\ 0, & otherwise \end{cases}. \tag{5.14}$$

User i updates the probability distribution $\pi_i(n) = \left(\pi_i^{(1)}(n), \cdots, \pi_i^{(K)}(n)\right)$ according to $v_i(n-1)$, and then chooses an action $a_i(n)$. Notice that the RS utilizes all the historical ratings of a user to compute recommendations. Even if the user does not rate any item in this iteration, the RS can still compute recommendations for him based on the ratings that the user has provided in previous iterations. Therefore, we use $a_i(n)$ to denote all the items that user i has rated by the end of iteration n, and naturally we have $a_i(n) \supseteq a_i(n-1)$.

If $v_i(n-1) = 0$, then user i may: (i) choose more items to rate, if he believes it is because he did not provide enough ratings that the recommendation result is unsatisfactory; (ii) keep previous action, i.e. rate no more items, if he blames the unsatisfying result on other users. For any $k \in \{1, \cdots, K\}$, $\pi_i^{(k)}(n) \triangleq \Pr\left(a_i(n) = A^{(k)}\right)$ is computed as follows:

$$\pi_i^{(k)}(n) = \begin{cases} \sigma_i(n-1), & if \quad A^{(k)} = a_i(n-1) \\ \beta_i(n)/\alpha^{c_i\left(A^{(k)}\right)}, & if \quad a_i(n-1) \subset A^{(k)} \subseteq S_i, \\ 0, & otherwise \end{cases} \tag{5.15}$$

where $\sigma_i (n - 1)$ is the rating completeness of user i:

$$\sigma_i (n - 1) = \frac{|a_i (n - 1)|}{|S_i|} . \tag{5.16}$$

A large $\sigma_i (n - 1)$ means user i has already rated many items in S_i, thus the user possibly rates no more items even if he is not satisfied with current recommendation. The normalization factor $\beta_i (n)$ is defined as follows:

$$\beta_i (n) = \frac{1 - \sigma_i (n - 1)}{\sum_{k: \, a_i (n-1) \subset A^{(k)} \subseteq S_i} \alpha^{-c_i (A^{(k)})}} . \tag{5.17}$$

If $v_i (n - 1) = 1$, then it is very likely that user i no longer rates the rest items in S_i. For any $k \in \{1, \cdots , K\}$, $\pi_i^{(k)} (n)$ is now defined as follows:

$$\pi_i^{(k)} (n) = \begin{cases} \mu, \; if \; A^{(k)} = a_i (n - 1) \\ \beta_i (n)/\alpha^{c_i (A^{(k)})}, \; if \; a_i (n - 1) \subset A^{(k)} \subseteq S_i , \\ 0, \; otherwise \end{cases} \tag{5.18}$$

where the parameter μ denotes to what extent a satisfied user would keep the previous action, and usually there is $0.5 < \mu \leq 1$. The normalization factor $\beta_i (n)$ is defined as follows:

$$\beta_i (n) = \frac{1 - \mu}{\sum_{k: \, a_i (n-1) \subset A^{(k)} \subseteq S_i} \alpha^{-c_i (A^{(k)})}} . \tag{5.19}$$

After every user has chosen his action, the RS computes the recommendations based on the rating matrix $\mathbf{R} (n)$ and returns $\hat{\mathbf{r}}_i (n)$ to user i. The learning process goes to the next iteration. If after a finite number of iterations, say n_s, all users have been satisfied, then the process stops. We say the behavior rule converges to an SE $\mathbf{a}^+ = (a_1 (n_s) , \cdots , a_N (n_s))$. A summary of the learning process is shown in Algorithm 1.

5.6 Convergence of the SE Learning Algorithm

In this section, we study the convergence of learning algorithm proposed in the previous section. First, we introduce the basic assumption for the convergence analysis and the notion of *user state*. Then we present a simple analysis of the convergence. After that, we make some simplifications of the learning algorithm and present a quantitative analysis of the convergence.

Algorithm 1 Learning the SE of the Game $\hat{G}_{CF} = \left(\mathcal{N}, \{\mathscr{A}_i\}_{i \in \mathcal{N}}, \{f_i\}_{i \in \mathcal{N}} \right)$

1: $n = 0$;

2: $\forall k \in \{1, \cdots, K\}$,

$$\pi_i^{(k)}(0) = \begin{cases} \beta_i(0)/\alpha^{c_i\left(A^{(k)}\right)}, & if \ A^{(k)} \subseteq S_i \\ 0, & otherwise \end{cases},$$

where $\beta_i(0) = \dfrac{1}{\sum\limits_{k: \ A^{(k)} \subseteq S_i} \alpha^{-c_i\left(A^{(k)}\right)}}$.

3: $a_i(0) \sim \pi_i(0)$;

4: **for all** $n > 0$ **do**

5:     update $\pi_i(n)$: $\forall k \in \{1, \cdots, K\}$,

$$\pi_i^{(k)}(n) = \begin{cases} \gamma_i(n), & if \ A^{(k)} = a_i(n-1) \\ \beta_i(n)/\alpha^{c_i\left(A^{(k)}\right)}, & if \ a_i(n-1) \subset A^{(k)} \subseteq S_i, \\ 0, & otherwise \end{cases} \text{where}$$

$$\gamma_i(n) = \begin{cases} \sigma_i(n-1), & if \ v_i(n-1) = 0 \\ \mu, & if \ v_i(n-1) = 1 \end{cases},$$

$$\beta_i(n) = \dfrac{1-\gamma_i(n)}{\sum\limits_{k: \ a_i(n-1) \subset A^{(k)} \subseteq S_i} \alpha^{-c_i\left(A^{(k)}\right)}}.$$

6:     $a_i(n) \sim \pi_i(n)$;

7: **end for**

5.6.1   Basic Assumption

The learning algorithm proposed above implies the following assumption we make about the relationship between the rating completeness and the recommendation quality:

Assumption 1 $\forall i \in \mathcal{N}$, the following two conditions hold for all $\sigma_i \in (0, 1]$ and $\sigma_{-i} \in (0, 1]$:

(i) $\frac{\partial h(\sigma_i, \sigma_{-i}; \mathbf{p}_i)}{\partial \sigma_i} > 0$;

(ii) $\frac{\partial h(\sigma_i, \sigma_{-i}; \mathbf{p}_i)}{\partial \sigma_{-i}} \geq 0$.

This assumption indicates that the recommendation quality perceived by one user can be improved by either the user himself or other users. During the learning process, unsatisfied users continually provide more ratings. For an unsatisfied user i, even if σ_{-i} no longer increases, so long as σ_i increases with iterations, the recommendation quality gradually improves, and the user may get a satisfying result after a few iterations. If the assumption doesn't hold, then it is impossible for the learning algorithm to achieve an SE if someone is unsatisfied with the initial recommendation. And in such an case, rating more items only increases the user's cost, and the user will lose his motivation to participate. Notice that the function $h(\cdot, \cdot; \mathbf{p}_i)$ is actually determined by the recommendation algorithm and the evaluation metric of recommendation quality. Hence we assume that given the evaluation metric of recommendation quality, the recommendation algorithm adopted by the RS should satisfy the above assumption. We have demonstrated the rationality of the assumption via simulation results on real data set (see Sect. 5.7.3).

5.6.2 User State

Based on Assumption 1, given a user's expectation Γ_i, the relationship between σ_i and σ_{-i} can be depicted by a curved section in the σ_i-σ_{-i} plane. As shown in Fig. 5.3, only when both σ_i and σ_{-i} exceed the corresponding thresholds, it is possible that user i will be satisfied. The two thresholds $\sigma_{i,\min}$ and $\sigma_{-i,\min}$ are determined by the following two equations respectively:

$$h\left(\sigma_{i,\min}, 1; \mathbf{p}_i\right) = \Gamma_i , \tag{5.20}$$

$$h\left(1, \sigma_{-i,\min}; \mathbf{p}_i\right) = \Gamma_i . \tag{5.21}$$

According to Assumption 1, if $\sigma_i < \sigma_{i,\min}$, then for any $\sigma_{-i} \in (0, 1]$, there is

$$h\left(\sigma_i, \sigma_{-i}; \mathbf{p}_i\right) < h\left(\sigma_{i,\min}, \sigma_{-i}; \mathbf{p}_i\right) \leq h\left(\sigma_{i,\min}, 1; \mathbf{p}_i\right). \tag{5.22}$$

Similarly, if $\sigma_{-i} < \sigma_{-i,\min}$, then for any $\sigma_i \in (0, 1]$, there is

$$h\left(\sigma_i, \sigma_{-i}; \mathbf{p}_i\right) < h\left(\sigma_i, \sigma_{-i,\min}; \mathbf{p}_i\right) \leq h\left(1, \sigma_{-i,\min}; \mathbf{p}_i\right). \tag{5.23}$$

Therefore, given Γ_i, $\sigma_{i,\min}$ represents the minimum requirement for user i and $\sigma_{-i,\min}$ represents the minimum requirement for other users.

During the learning process, each user's rating completeness increases with iterations. We define $\sigma_i (n - 1)$ and $\sigma_{-i} (n - 1)$ as follows:

$$\sigma_i (n - 1) = \frac{|a_i (n - 1)|}{|S_i|} , \tag{5.24}$$

$$\sigma_{-i} (n - 1) = \frac{1}{N - 1} \sum_{j \in \mathcal{N}, \, j \neq i} \frac{|a_j (n - 1)|}{|S_j|} . \tag{5.25}$$

For any $n \geq 1$, there is $\sigma_i (n) \geq \sigma_i (n - 1)$ and $\sigma_{-i} (n) \geq \sigma_{-i} (n - 1)$. We assume that there exists some n_0 ($n_0 \geq 1$) that $\sigma_i (n_0) \geq \sigma_{i,\min}$ holds for all i. From iteration $n_0 + 1$, each user i is in one of the following three states:

- *Satisfied*: as depicted by the green area in Fig. 5.3, user i has already got satisfying recommendations, namely $h\left(\sigma_i (n - 1), \sigma_{-i} (n - 1); \mathbf{p}_i\right) \geq \Gamma_i$. Once the user is satisfied, he will always in the *Satisfied* state. The reason is that with the increase of iterations, σ_i and σ_{-i} either increase or remain the same, and according to *Assumption 1*, $h\left(\sigma_i, \sigma_{-i}; \mathbf{p}_i\right)$ will not decrease.

Fig. 5.3 Illustration of user
state: *Satisfied* (cyan area),
Proximity to satisfied (blue
area), *Far from satisfied*
(yellow area)

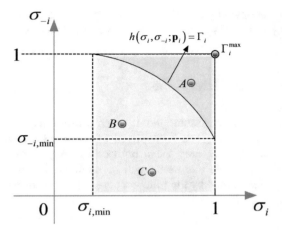

- *Proximity to satisfied*: as depicted by the cyan area in Fig. 5.3, user i has not been satisfied, namely $h(\sigma_i(n-1), \sigma_{-i}(n-1); \mathbf{p}_i) < \Gamma_i$, and user i has not rated all the items in S_i, namely $\sigma_i(n-1) < 1$, while other users have rated enough items, namely $\sigma_{-i}(n-1) \geq \sigma_{-i,\min}$. In this case, even if other users no longer rate more items, user i is able to enter the *Satisfied* state by rating more items.
- *Far from satisfied*: as depicted by the yellow area in Fig. 5.3, user i has not been satisfied, and the amount of ratings provided by other users has not achieved the minimum requirement of user i, namely $\sigma_{-i}(n-1) < \sigma_{-i,\min}$. In this case, if other users provide enough ratings in subsequent iterations, user i can enter the *Proximity to satisfied* state. Otherwise, the user will stuck in this state and never be satisfied.

We use Z_S, Z_P and Z_F to denote the three states respectively, and we use $z_i(n)$ to denote user i's state at the beginning of iteration n $(n \geq n_0)$, then $z_i(n) \in \{Z_S, Z_P, Z_F\}$.

5.6.3 Simple Analysis of the Convergence

At the beginning of iteration n $(n \geq n_0)$, users can be grouped into two sets: the set of satisfied users $\mathcal{N}_S(n) \triangleq \{i | i \in \mathcal{N}, z_i(n) = Z_S\}$, and the set of unsatisfied users $\mathcal{N}_{US}(n) \triangleq \{i | i \in \mathcal{N}, z_i(n) = Z_P \vee z_i(n) = Z_F\}$. As the learning process continues, the number of unsatisfied users decreases. For any user $i \in \mathcal{N}_{US}(n)$:

(i) If $z_i(n) = Z_P$, then according to the definition of the state Z_P, the user will eventually become satisfied.

(ii) If $z_i(n) = Z_F$, there is $\sigma_{-i}(n-1) < \sigma_{-i,\min}$. When users continue to provide more ratings after they are satisfied, namely $\mu < 1$, both $\sigma_{-i}(n-1)$ and $\sigma_i(n-1)$ keep increasing with n, hence at some iteration n' $(n' > n)$, there will be $z_i\left(n'\right) = Z_P$. However, when users in $\mathcal{N}_S\left(n\right)$ make no contributions to the increase of $\sigma_{-i}\left(n-1\right)$, namely $\mu = 1$, it is possible that the user will stay in the state Z_F permanently. Consider the case that the satisfied users have provided so few ratings that σ_{-i} cannot reach $\sigma_{-i,\min}$ even when all the unsatisfied users provide their complete ratings. In such a case, user i will never be satisfied.

To sum up, given $\mu = 1$, if the following inequality holds for some $i \in \mathcal{N}$ and some $n \in \{1, 2, \cdots\}$, then the learning algorithm cannot converge:

$$\frac{1}{N-1}\left[\sum_{j\in\mathcal{N}_S(n)}\sigma_j\left(n-1\right) + \sum_{j\in\mathcal{N}_{US}(n),\ j\neq i}1\right] < \sigma_{-i,\min} \tag{5.26}$$

Notice that $\sigma_{-i,\min}$ is determined by user i's expectation Γ_i. According to (5.26), we can conclude that if a *small* portion of users have relatively *high* expectations for the recommendation quality, then the proposed learning algorithm cannot converge to an SE. Next we will present an elaborate analysis of this conclusion.

5.6.4 Quantitative Analysis of the Convergence

From above discussion we can see that, to judge the convergence of the learning algorithm, the key is to analyze how each user's rating completeness changes over time. Let $\Delta\sigma_i\left(n\right)$ denote the increment of user i's rating completeness in iteration n, namely $\Delta\sigma_i\left(n\right) = \sigma_i\left(n\right) - \sigma_i\left(n-1\right)$. According to Algorithm 1, the value of $\Delta\sigma_i\left(n\right)$ is random. Hence it is difficult to quantitatively analyze the transition of user state. In this section, we simplify the learning process described in Algorithm 1 and discuss the convergence of the simplified version.

5.6.4.1 Simplified Learning Algorithm

The simplified learning process can be described as follows. Initially, each user i randomly chooses one item from S_i, thus for any $i \in \mathcal{N}$, $\sigma_i\left(0\right) = \frac{1}{|S_i|}$. At the beginning of iteration n ($n > 0$), each user i judges whether the recommendation quality is satisfactory. If the user is satisfied, then he doesn't change the action, namely $a_i\left(n\right) = a_i\left(n-1\right)$; if the user is unsatisfied, then he randomly chooses one item from $S_i \backslash a_i\left(n-1\right)$.

Based on above simplification, we can easily determine the value of $\Delta\sigma_i\left(n\right)$: if $i \in \mathcal{N}_S\left(n\right)$, then $\Delta\sigma_i\left(n\right) = 0$; if $i \in \mathcal{N}_{US}\left(n\right)$, then $\Delta\sigma_i\left(n\right) = \frac{1}{|S_i|}$.

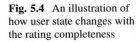
Fig. 5.4 An illustration of how user state changes with the rating completeness

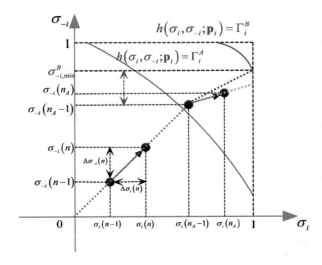

5.6.4.2 Two Types of Users

In addition to simplifying the learning process, we also make some assumptions about users:

Assumption 2 Users in $\mathcal{N}$ can be divided into two groups $\mathcal{N}_A$ and $\mathcal{N}_B$.

(i) For all $i \in \mathcal{N}_A$, $\Gamma_i = \eta_A \Gamma_i^{\max}$, where $0 < \eta_A < 1$.
(ii) For all $i \in \mathcal{N}_B$, $\Gamma_i = \eta_B \Gamma_i^{\max}$, where $\eta_A < \eta_B < 1$.
(iii) For all $i \in \mathcal{N}$, $|S_i| = M_0$, where M_0 is constant and $1 \leq M_0 < |S|$.

5.6.4.3 Quantify the Change of Rating Completeness

With above simplifications, we can now quantitatively analyze when the transition of user state happens and explain why users in $\mathcal{N}_B$ may stay unsatisfied. Consider a user $i \in \mathcal{N}_B$. As shown in Fig. 5.4, with the increase of iterations, the user "moves" upwards and/or rightwards in the square $[0, 1]^2$. To judge whether the user can enter the *Satisfied* area which is determined by η_B, we need to analyze how the user moves from $(\sigma_i (n - 1), \sigma_{-i} (n - 1))$ to $(\sigma_i (n), \sigma_{-i} (n))$ in each iteration n.

According to the simplified algorithm, at the beginning of iteration 1, user i is at the point $\left(\frac{1}{M_0}, \frac{1}{M_0} \right)$. Then, the user moves along the line $\sigma_{-i} = \sigma_i$ (the blue dotted line in Fig. 5.4) until users in $\mathcal{N}_A$ are satisfied. This is because before anyone is satisfied, for all $i \in \mathcal{N}$, both $\Delta\sigma_i (n)$ and $\Delta\sigma_{-i} (n)$ equal to $\frac{1}{M_0}$. Users in $\mathcal{N}_A$ have relatively low expectations, thus they become satisfied earlier than users in $\mathcal{N}_B$. There exists some $n_A \in \{1, 2, \ldots, M_0 - 1\}$ that at the beginning of iteration n_A, all the users in $\mathcal{N}_A$ are satisfied, while all the users in $\mathcal{N}_B$ are unsatisfied.

At the beginning of iteration n_A, user i is at the point $\left(\frac{n_A}{M_0}, \frac{n_A}{M_0}\right)$. As we have discussed in Sect. 5.6.3, if user i is in the state Z_P, then he can be satisfied by rating more items. Here we focus on the other case, namely $z_i(n_A) = Z_P$. During iteration n_A, each user in $\mathcal{N}_B$ rates one more item, while users in $\mathcal{N}_A$ no longer rate more items, hence user i moves along a line whose slope k_B is less than 1:

$$
\begin{aligned}
k_B &= \frac{\Delta\sigma_{-i}(n_A)}{\Delta\sigma_i(n_A)} \\[2mm]
&= \frac{\frac{1}{N-1}\sum\limits_{j\in\mathcal{N}_B, j\neq i}\left[\sigma_j(n_A) - \sigma_j(n_A - 1)\right]}{\frac{1}{M_0}} \\[2mm]
&= \frac{|\mathcal{N}_B| - 1}{N - 1}.
\end{aligned}
\tag{5.27}
$$

In subsequent iterations, user i moves along the same direction until one of the following two situations happens: (i) user i becomes satisfied; (ii) user i hasn't been satisfied but has rated all the items in S_i, namely $\sigma_i = 1$. If the second situation happens, user i can never be satisfied. This is because that users in $\mathcal{N}_B$ are assumed to have same expectations, which implies all the other users in $\mathcal{N}_B$ also have provided their complete ratings. As a result, no user can make contributions to the increase of σ_{-i}. As depicted by the green dotted line in Fig. 5.4, the second situation will happen if k_B is smaller than some threshold k_{min} (see the red dotted line in Fig. 5.4):

$$
k_{min} = \frac{\sigma_{-i,min} - \sigma_{-i}(n_A - 1)}{1 - \sigma_i(n_A - 1)} = \frac{\sigma_{-i,min} - \frac{n_A}{M_0}}{1 - \frac{n_A}{M_0}}.
\tag{5.28}
$$

Plugging (5.27) and (5.28) into $k_B < k_{min}$ we can get:

$$
\sigma_{-i,min} > \frac{|\mathcal{N}_B| - 1}{N - 1} + \frac{N - |\mathcal{N}_B|}{N - 1}\cdot\frac{n_A}{M_0}.
\tag{5.29}
$$

The right part of above inequality is exactly the formula for calculating σ_{-i} when users in $\mathcal{N}_B$ provide their complete ratings and users in $\mathcal{N}_A$ provide the necessary amount of ratings to make themselves satisfied. Similar with the conclusion we've drawn in Sect. 5.6.3, the inequality implies that if a user has very high expectation which requires too much effort of other users, then the user cannot get satisfying recommendations. If the relationship between σ_i, σ_{-i}, η_A, and η_B can be explicitly expressed, then we can rewrite $\sigma_{-i,min}$ in a specific form, and the influence of users' expectations on the convergence of the learning algorithm can be shown more clearly. Next we'll show how to quantify the influence.

Fig. 5.5 The relationship between σ_i and σ_{-i} with respect to a given Γ_i

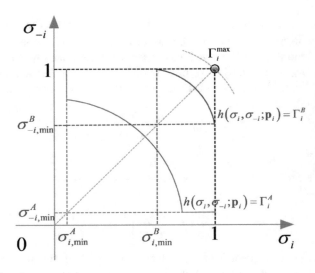

5.6.5 Convergence Conditions of the Simplified Learning Algorithm

To derive specific convergence conditions from (5.29), we make the following assumption as a complementary to Assumption 2: for all $i \in \mathcal{N}$, given $\eta_i \in \{\eta_A, \eta_B\}$, the relationship between σ_i and σ_{-i} can be formulated as:

$$\sigma_i^2 + \sigma_{-i}^2 = 2\eta_i^2, \tag{5.30}$$

where $\frac{1}{M_0} \le \sigma_i \le 1$, $\frac{1}{M_0} \le \sigma_{-i} \le 1$.

In above assumption, the quadratic relationship between σ_i and σ_{-i} is proposed based on simulation results on real data set (see Sect. 5.7.3). As shown in Fig. 5.5, the threshold $\sigma_{-i,\min}$ is now defined in the following way:

(i) If $\eta_A < \eta_B \le \frac{1}{\sqrt{2}}$, then for all $i \in \mathcal{N}_B$, $\sigma_{-i,\min} = \frac{1}{M_0}$. Considering that $n_A \ge 1$, (5.29) implies that:

$$\frac{1}{M_0} > \frac{|\mathcal{N}_B| - 1}{N - 1} + \frac{N - |\mathcal{N}_B|}{N - 1} \cdot \frac{1}{M_0}. \tag{5.31}$$

Then we can get:

$$(M_0 - 1)\,(|\mathcal{N}_B| - 1) < 0 \tag{5.32}$$

Because $M_0 \ge 1$ and $|\mathcal{N}_B| \ge 1$, above inequality doesn't hold. Therefore, when $\eta_A < \eta_B \le \frac{1}{\sqrt{2}}$, it is impossible that $k_B < k_{\min}$, which means the algorithm must converge.

(ii) If $\frac{1}{\sqrt{2}} < \eta_B < 1$, then for all $i \in \mathcal{N}_B$, there is:

$$\sigma_{-i,\min} = \sqrt{2\eta_B^2 - 1}. \tag{5.33}$$

From $z_i(n_A) = Z_F$ we can get:

$$\frac{n_A}{M_0} < \sqrt{2\eta_B^2 - 1}. \tag{5.34}$$

On the other hand, for any user $i \in \mathcal{N}_A$, the following inequality holds:

$$[\sigma_i(n_A - 1)]^2 + [\sigma_{-i}(n_A - 1)]^2 \geq 2\eta_A^2, \tag{5.35}$$

where $\sigma_i(n_A - 1) = \sigma_{-i}(n_A - 1) = \frac{n_A}{M_0}$, then we get:

$$\frac{n_A}{M_0} \geq \eta_A. \tag{5.36}$$

From (5.29), (5.33), (5.34) and (5.36) we can get:

$$\sqrt{2\eta_B^2 - 1} > \frac{|\mathcal{N}_B| - 1}{N - 1} + \frac{N - |\mathcal{N}_B|}{N - 1}\eta_A. \tag{5.37}$$

Based on above discussions, we can provide the following proposition:

Proposition 1 *The simplified learning algorithm can converge to an SE of the game* $\hat{G}_{CF} = (\mathcal{N}, \{\mathcal{A}_i\}_{i \in \mathcal{N}}, \{f_i\}_{i \in \mathcal{N}})$ *if Assumption 2 holds and one of the following two conditions holds:*

(i) $\eta_A < \eta_B \leq \frac{1}{\sqrt{2}}$;
(ii) $\frac{1}{\sqrt{2}} < \eta_B < 1$ *and*

$$\sqrt{2\eta_B^2 - 1} \leq \frac{|\mathcal{N}_B| - 1}{N - 1} + \frac{N - |\mathcal{N}_B|}{N - 1}\eta_A. \tag{5.38}$$

To better understand the influence of η_A, η_B and $|\mathcal{N}_B|$ on the convergence of the learning algorithm, we make following discussions:

(i) Given $\rho_N \triangleq \frac{|\mathcal{N}_B|}{N}$, according to (5.37), the simplified algorithm cannot converge if the following condition holds:

$$\eta_B > \sqrt{\frac{1}{2}\left[\frac{\rho_N N - 1}{N - 1} + \frac{(1 - \rho_N) N}{N - 1}\eta_A\right]^2 + \frac{1}{2}}. \tag{5.39}$$

Fig. 5.6 Illustration of the relationship between θ_B and η_A. Given ρ_N and η_A, the simplified learning algorithm cannot converge to an SE if $\eta_B > \theta_B$. We set $0.3 \le \eta_A \le 0.9$ and $N = 10{,}000$ to compute θ_B

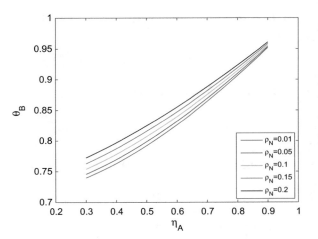

We use θ_B to denote the right side of (5.39). Figure 5.6 illustrates how θ_B changes with η_A under different settings of ρ_N. As we can see, given ρ_N, θ_B grows with η_A, but the growth rate of θ_B is lower than that of η_A. This means as expectations of most users become higher (larger η_A), even if users don't make significant difference on their expectations, there may be some users who can never be satisfied. From Fig. 5.6 we can also observe that for a given η_A, θ_B increases with ρ_N. This implies that as more users have high expectations (larger ρ_N), users can expect higher recommendation quality (larger η_B).

(ii) Given $\rho_\eta \triangleq \frac{\eta_B}{\eta_A}$, the simplified algorithm cannot converge if $\frac{1}{\sqrt{2}} < \eta_B < 1$ and

$$\frac{|\mathcal{N}_B|}{N} < \frac{(N-1)\sqrt{2\eta_B^2 - 1} + 1 - N\frac{\eta_B}{\rho_\eta}}{N\left(1 - \frac{\eta_B}{\rho_\eta}\right)}. \tag{5.40}$$

We use θ_N to denote the right side of above inequality. The inequality implies $\theta_N > 0$ from which the following two conditions can be derived:

$$N > \frac{\sqrt{2}}{\sqrt{2} - \frac{1}{\rho_\eta}}, \tag{5.41}$$

$$\eta_B > \frac{-\frac{2N}{\rho_\eta} + \sqrt{\Delta_f}}{2\left[2(N-1)^2 - \frac{N^2}{\rho_\eta^2}\right]}, \tag{5.42}$$

where $\Delta_f = \left(\frac{2N}{\rho_\eta}\right)^2 + 4\left[2(N-1)^2 - \frac{N^2}{\rho_\eta^2}\right](N^2 - 2N + 2)$. We use $\eta_{B,\min}$ to denote the right side of (5.42). Figure 5.7 illustrates how θ_N changes with η_B.

Fig. 5.7 Illustration of the relationship between θ_N and η_B. Dotted line marks $\eta_{B,min}$ corresponding to a given ρ_η. Given ρ_η, the simplified learning algorithm cannot converge to an SE if $\eta_B > \eta_{B,min}$ and $\rho_N < \theta_N$. We set $\frac{1}{\sqrt{2}} < \eta_B \le 0.9$ and $N = 10,000$ to compute θ_N

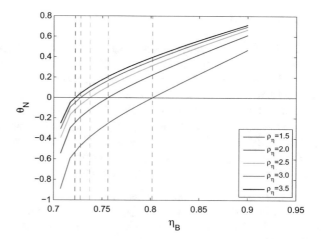

As we can see, given ρ_η, θ_N increases with η_B, which implies when users with high expectation expect higher recommendation quality, there should be more such users so that an SE can be achieved. Given η_B, θ_N increases with ρ_η, which implies that as the difference on expected recommendation quality between two types of users becomes wider, more users can have high expectations. These implications are consistent with those we get from Fig. 5.6.

5.7 Simulation

To verify the feasibility of the proposed SE learning algorithm, we have conducted a series of simulations by using real rating data. In this section, we first describe the preparation of data and experiment setup, then we present a comparison of the learning results which are obtained under different settings of users' expectations. After that, we provide some experimental proofs for the assumptions we've made and for the theoretical analysis presented in Sect. 5.6. In addition, to demonstrate the satisfactory game analysis can help to design incentive mechanisms for user participation, we conduct experiments to investigate how monetary rewards affect the result of equilibrium learning.

5.7.1 Data Set and Parameter Setting

Two data sets, namely Jester [19] and MovieLens,[2] are chosen for simulation. These two data sets are commonly used in the study of collaborative filtering. Details of the data sets and corresponding parameter settings are given below.

---

[2]http://grouplens.org/datasets/movielens/1m/.

5.7.1.1 Jester

The Jester data set contains about 4.1 million ratings of 100 jokes from 73,421 users. Considering that the "ground truth" of a user's preference for each item is required for the evaluation of recommendation quality, we only keep 720,000 ratings from the 7200 users who have rated all the 100 jokes. Ratings are real values ranging from -10.00 to $+10.00$ (the value "99" corresponds to "unrated"). As described in Sect. 5.3, we have defined $0 \leq r_{ij} \leq r_{\max}$, so we adjust the ratings to the range [10.00, 30.00] and use "0" to represent "not rated". Finally we get a user-item matrix $\mathbf{R} = [r_{ij}]_{7200 \times 100}$ which contains no zero elements.

Parameters of the SE learning algorithm are set as follows:

- $\mathbf{p}_i$: Each row of $\mathbf{R}$ is treated as the corresponding user's interest vector.
- S_i: For each user i, we set $|S_i| = 70$ and randomly set 30% of the user's ratings to "0". The resulting rating matrix is denoted by $\mathbf{R}'$.
- $g_i(\hat{\mathbf{r}}_i)$: The quality of the recommendation $\hat{\mathbf{r}}_i$ is evaluated according to (5.4).
- $c_i(a_i)$: The rating cost is defined as the number of rated items, that is, $c_i(a_i) = |a_i|$.
- α: This parameter affects the convergence speed. Considering the shape of the function $f(x) = 1/\alpha^x$ on the interval $[0, |S_i|]$, we set $\alpha = 1.2$.
- μ: We set $\mu = 0.9$ and $\mu = 1$ to simulate the situation that satisfied users continue to provide ratings and the situation that satisfied users no longer rate more items respectively.

We have implemented a user-based CF algorithm with MATLAB. Unknown ratings are predicted according to (5.2) where $|Neighbour(i)|$ is set to 36. The quality of recommendations is evaluated according to (5.4). Based on $\mathbf{R}$ and $\mathbf{R}'$, we calculate the best result $\Gamma_i^{\max}$, then we set $\Gamma_i = \eta_i \Gamma_i^{\max}$, where $0 < \eta_i < 1$. Settings of $\{\eta_i\}_{i=1}^N$ will be described later.

5.7.1.2 MovieLens

The MovieLens data set contains about one million ratings from 6040 users on 3900 movies. To conduct simulations, we set $|S_i| = 70$ and discard users who rated less than 70 movies. The resulting data set consists of ratings from 3631 users on 3675 movies. Let $\mathbf{R}' = [r_{ij}]_{3631 \times 3675}$ denote the rating matrix, where $r_{ij} \in \{0, 1, \cdots, 5\}$ and $r_{ij} = 0$ means "unrated". This rating matrix is quite sparse: the proportion of non-zero elements is only 6.78%. To determine the interest vector $\mathbf{p}_i$ of each user, we first apply the CF algorithm to predict the unknown ratings in $\mathbf{R}'$. Let $\mathbf{R}$ denote the matrix which consists of original ratings and predicted ratings. Then each row of $\mathbf{R}$ is treated as the corresponding user's interest vector. Other parameters of the SE learning algorithm are set in the same way as the Jester data.

Table 5.1 Simulation results of satisfaction equilibrium learning on Jester data set

		$\mu = 0.9$					$\mu = 1$				
	runID	1	2	3	4	5	1	2	3	4	5
$\eta_i = 0.5$	n_{stop}	24	28	20	22	26	158	84	130	257	151
	N_S	7200	7200	7200	7200	7200	7200	7200	7200	7200	7200
	$\bar{\sigma}_i$	0.531	0.553	0.507	0.519	0.543	0.280	0.281	0.282	0.282	0.282
$\eta_i = 0.85$	n_{stop}	1203	1307	1306	1206	1203	8893	9560	9813	8988	9560
	N_S	7200	7200	7200	7200	7200	7200	7200	7200	7200	7200
	$\bar{\sigma}_i$	0.913	0.936	0.931	0.928	0.914	0.781	0.782	0.781	0.782	0.782
1%: $\eta_i = 0.85$	n_{stop}	406	404	504	504	403	10,000	10,000	10,000	10,000	10,000
99%: $\eta_i = 0.5$	N_S	7200	7200	7200	7200	7200	7130	7130	7129	7130	7131
	$\bar{\sigma}_i$	0.906	0.888	0.895	0.896	0.881	0.325	0.326	0.326	0.324	0.324
20%: $\eta_i = 0.85$	n_{stop}	804	905	802	902	903	10,000	10,000	10,000	10,000	10,000
80%: $\eta_i = 0.5$	N_S	7200	7200	7200	7200	7200	6901	6899	6885	6891	6896
	$\bar{\sigma}_i$	0.911	0.920	0.897	0.903	0.910	0.441	0.441	0.441	0.442	0.440

5.7.2 Simulation Results of SE Learning

To verify the convergence of Algorithm 1, we test multiple groups of $\{\eta_i\}_{i=1}^N$. For a given μ, we run simulations with the following four settings: (i) $\eta_i = 0.5$ for all $i \in \mathcal{N}$; (ii) $\eta_i = 0.85$ for all $i \in \mathcal{N}$; (iii) $\eta_i = 0.85$ for 1% of the users, $\eta_i = 0.5$ for the rest; (iv) $\eta_i = 0.85$ for 20% of the users, $\eta_i = 0.5$ for the rest. To reduce the influence of randomness, we run the algorithm five times for each setting. In each run, the iterative process stops when all users are satisfied or the number of iterations reaches 10,000. After each run, we record the number of iterations n_{stop}, the number of satisfied users N_S, and the average rating completeness $\bar{\sigma}_i \triangleq \frac{1}{N} \sum_{i=1}^N |a_i\left(n_{stop}\right)|/|S_i|$. Simulation results are shown in Tables 5.1 and 5.2, from which we can make the following observations.

When users have similar expectations for the recommendation quality, even if the expectation is high ($\eta_i = 0.85$) and user becomes inactive after he is satisfied ($\mu = 1$), an SE can be reached. For a given μ, as users' expectations become higher, the convergence time becomes longer, and $\bar{\sigma}_i$ becomes higher, which means users need to rate more items. Given the setting of η_i, by comparing the results of different μ we can see that, when satisfied users no longer rate more items, the convergence time becomes longer, while the average rating completeness decreases. For example, as shown in Table 5.1, given $\eta_i = 0.5$ for all $i \in \mathcal{N}$, when $\mu = 0.9$, an SE can be reached in 30 iterations, and averagely a user needs to rate 50–60% of the items that he has experienced; when $\mu = 1$, usually more than 100 iterations are required to reach an SE, while the user only needs to rate less than 30% of the items. During the learning process, due to the randomness of users' actions, different users become satisfied at different time. If $\mu = 0.9$, satisfied users continue to make contributions to the improvement of recommendation quality, hence those unsatisfied users can

Table 5.2 Simulation results of satisfaction equilibrium learning on MovieLens data set

runID	$\mu = 0.9$					$\mu = 1$				
	1	2	3	4	5	1	2	3	4	5
$\eta_i = 0.5$										
n_{stop}	27	29	48	28	73	57	28	39	53	37
N_S	3631	3631	3631	3631	3631	3631	3631	3631	3631	3631
$\bar{\sigma}_i$	0.814	0.824	0.884	0.821	0.921	0.224	0.224	0.223	0.223	0.226
$\eta_i = 0.85$										
n_{stop}	528	679	533	494	560	734	610	640	780	571
N_S	3631	3631	3631	3631	3631	3631	3631	3631	3631	3631
$\bar{\sigma}_i$	0.990	0.994	0.990	0.989	0.991	0.745	0.744	0.743	0.745	0.745
$1\%: \eta_i = 0.85$ **$99\%: \eta_i = 0.5$**										
n_{stop}	118	78	223	68	157	10,000	10,000	10,000	10,000	10,000
N_S	3631	3631	3631	3631	3631	3630	3629	3629	3628	3628
$\bar{\sigma}_i$	0.952	0.928	0.975	0.919	0.964	0.234	0.231	0.229	0.231	0.229
$20\%: \eta_i = 0.85$ **$80\%: \eta_i = 0.5$**										
n_{stop}	482	458	302	462	678	10,000	10,000	10,000	10,000	10,000
N_S	3631	3631	3631	3631	3631	3624	3619	3625	3622	3624
$\bar{\sigma}_i$	0.989	0.989	0.981	0.989	0.993	0.350	0.349	0.350	0.349	0.350

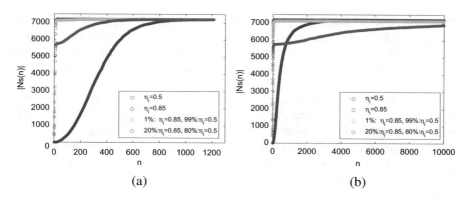

Fig. 5.8 The change of the number of satisfied users. Simulations are conducted on Jester data set. (**a**) $\mu = 0.9$. (**b**) $\mu = 1$

be satisfied in a short time. While if $\mu = 1$, unsatisfied users can only rely on themselves to improve the recommendation quality, hence more time is needed. As for the rating completeness, $\mu = 0.9$ means the user may provide more ratings after he is satisfied, thus by the time an SE is reached, the user may have rated much more items than he needs to. While $\mu = 1$ means the user prefers to rate the minimum number of items necessary to get satisfactory recommendations, thus when an SE is achieved, the average rating completeness is lower than that of $\mu = 0.9$.

When most users have moderate expectations for the recommendation quality ($\eta_i = 0.5$) and a small portion of users have much higher expectations ($\eta_i = 0.85$), an SE can still be reached when $\mu = 0.9$, although the convergence time is much longer than that when all users have moderate expectations, and the average rating completeness is close to that when all users have high expectations. This result implies that in order to meet the high expectations of a few users, users with moderate expectations have to rate much more items after they are satisfied. When $\mu = 1$, satisfied users no longer rate more items. Hence, after the majority of users have been satisfied, those unsatisfied users can hardly improve the recommendation quality. For example, as shown in Table 5.1, under the third setting of η_i, the 1% of users who have high expectations are still unsatisfied after 10,000 iterations. From the corresponding $\bar{\sigma}_i$ we can learn that most users just rate "enough" number of items to meet their moderate expectations, while the amount of their ratings is far from enough to achieve the expectations of the rest 72 users.

To better understand the influence of the minority high expectations on the learning results, we take a detailed look at the results on Jester data set and draw the sets of $|\mathcal{N}_S(n)|$ corresponding to different settings in Fig. 5.8. As shown in Fig. 5.8a, in a setting where $\eta_i = 0.85$ for 20% of the users (depicted by magenta circles), after 15 iterations, nearly 80% of the users are already satisfied. During the first 15 iterations, $|\mathcal{N}_S(n)|$ grows at almost the same rate with that of the setting where $\eta_i = 0.5$ for all users (depicted by red circles). After most users are satisfied, the growth rate of $|\mathcal{N}_S(n)|$ decreases. This is because the satisfied users prefer rating no more items, and for those unsatisfied users the recommendation results only improve a little after one iteration. Consequently, many more iterations are required to achieve the expectations of the rest users. Similar results can be observed in Fig. 5.8b.

The simulation results coincide with our intuition about the satisfaction equilibrium in a CF system: when all users have moderate expectations for recommendation quality, an SE can be realized in low cost, that is, every user can get satisfactory recommendations without rating many items. From the results shown in Tables 5.1 and 5.2 we can get some general insight about the convergence conditions of the learning algorithm. Next we will conduct another group of simulations to verify the analysis presented in Sect. 5.6.

5.7.3 Relationship Between Recommendation Quality and Rating Completeness

The theoretical analysis we presented in Sect. 5.6 is based on some assumptions (see *Assumptions 1* and *2*). Before we verify the convergence conditions, we first conduct some experiments on Jester data set to validate the rationality of the assumptions. For each user i, we utilize $\mathbf{R}'$ to construct a group of rating matrices $\{\mathbf{R}_{i,k}\}$. Each matrix $\mathbf{R}_{i,k}$ corresponds to a certain pair of σ_i and σ_{-i}, where $\sigma_i \in \left\{ \frac{1}{70}, \frac{2}{70}, \cdots, \frac{70}{70} \right\}$ and $\sigma_{-i} \in \left\{ \frac{1}{100}, \frac{2}{100}, \cdots, \frac{100}{100} \right\}$. For example, given $\sigma_i = \frac{5}{70}$ and $\sigma_{-i} = \frac{10}{100}$, we randomly choose five non-zero ratings from the ith row of $\mathbf{R}'$ and set them to 0, then we randomly set $\left(1 - \frac{10}{100}\right) \times 100\%$ of the non-zero ratings in other rows to 0. We apply the user-based CF algorithm (see (5.2)) to $\mathbf{R}_{i,k}$, and evaluate the recommendation results based on (5.4) to get $h(\sigma_i, \sigma_{-i}; \mathbf{p}_i)$. By drawing $\{(\sigma_i, \sigma_{-i}, h(\sigma_i, \sigma_{-i}; \mathbf{p}_i))\}$ in a three-dimensional space, we can get a plot of $h(\sigma_i, \sigma_{-i}; \mathbf{p}_i)$ corresponding to the user i. Figure 5.9 shows an example. From Fig. 5.9a we can see that the recommendation quality improves when σ_i or σ_{-i} increases. This result confirms *Assumption 1*. From the contour plot shown in Fig. 5.9b we can observe that, given a proper value of $h(\sigma_i, \sigma_{-i}; \mathbf{p}_i)$, there is an approximate quadratic relationship between σ_i and σ_{-i}. *Assumption 2* is proposed based on the this observation. Experiment results of other users can also support the two assumptions.

5.7.4 Convergence Test

We implement the simplified learning algorithm described in Sect. 5.6 and conduct simulations on Jester data set to verify the convergence conditions proposed in the

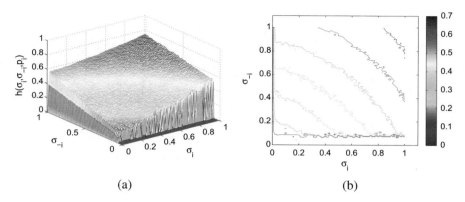

Fig. 5.9 Illustration of the relationship between recommendation quality and rating completeness: (a) the fitted surface is obtained by interpolating the experiment results $\{(\sigma_i, \sigma_{-i}, h(\sigma_i, \sigma_{-i}; \mathbf{p}_i))\}$ corresponding to one user; (b) the contour plot of (a)

appendix. As described in *Assumption 2*, we randomly divide users into two groups $\mathscr{N}_A$ and $\mathscr{N}_A$, and we set η_A to 0.5. The two parameters ρ_N and η_B are set in a following way: given $\rho_N \in \{0.01, 0.10, 0.20\}$, we calculate the corresponding θ_B according to (5.42), then we set $\eta_B = \theta_B - 0.05$, $\eta_B = \theta_B$ and $\eta_B = \theta_B + 0.05$ respectively. To prove that the algorithm can converge when all users have high expectations, we also test the setting $\eta_A = \eta_B = \theta_B + 0.05$. Given a group of (ρ_N, η_A, η_B), we run the simplified learning algorithm ten times. The iterative process stops at iteration n_{stop} when one of the following conditions is met: (i) $|\mathscr{N}_S(n_{stop})| = N$; (ii) $\forall i \in \mathscr{N}_{US}(n_{stop})$, $\sigma_i(n_{stop}) = 1$.

Let N_S denote the number of satisfied users at the end of the learning process, namely $N_S = |\mathscr{N}_S(n_{stop})|$. Table 5.3 shows the simulation results. As we can see, given η_A and ρ_N, an SE can always be reached when $\eta_B = \theta_B - 0.05$. When $\eta_B = \theta_B$, some users in $\mathscr{N}_B$ can be satisfied and some cannot. This result is slightly different from the theoretical analysis presented in Sect. 5.6, where (5.39) implies that an SE can be achieved when $\eta_B \leq \theta_B$. We think the reason for the inconsistence between theoretical analysis and simulation results is that the relationship between users' rating completeness and recommendation quality doesn't exactly accord with assumption we have made in (5.30). When $\eta_B = \theta_B + 0.05$, most of the users in $\mathscr{N}_B$ cannot be satisfied. From Table 5.3 we can also see that when users have similar high expectations, the simplified learning algorithm can converge to an SE before users have rated all the item they have experienced ($n_{stop} < |S_i|$). These results demonstrate that in a CF system, whether an equilibrium can be achieved via users' spontaneous participation depends on whether the users are homogeneous in the sense that they expect same recommendation quality.

Table 5.3 Simulations results of the simplified learning algorithm

ρ_N	η_A	θ_B	η_B	runID	1	2	3	4	5
0.01	0.5	0.792	0.742	n_{stop}	68	67	68	67	68
				N_S	7200	7200	7200	7200	7200
			0.792	n_{stop}	70	70	70	70	70
				N_S	7178	7178	7177	7176	7176
			0.842	n_{stop}	70	70	70	70	70
				N_S	7130	7130	7131	7130	7129
$\eta_A = \eta_B = 0.842$				n_{stop}	65	65	65	65	65
				N_S	7200	7200	7200	7200	7200
0.1	0.5	0.807	0.757	n_{stop}	69	68	68	68	69
				N_S	7200	7200	7200	7200	7200
			0.807	n_{stop}	70	70	70	70	70
				N_S	7132	7130	7139	7141	7141
			0.857	n_{stop}	70	70	70	70	70
				N_S	6642	6618	6627	6630	6619
$\eta_A = \eta_B = 0.857$				n_{stop}	66	66	65	66	66
				N_S	7200	7200	7200	7200	7200
0.2	0.5	0.824	0.775	n_{stop}	69	68	68	69	69
				N_S	7200	7200	7200	7200	7200
			0.825	n_{stop}	70	70	70	70	70
				N_S	7133	7146	7139	7142	7130
			0.875	n_{stop}	70	70	70	70	70
				N_S	6368	6400	6383	6398	6387
$\eta_A = \eta_B = 0.875$				n_{stop}	66	67	67	66	67
				N_S	7200	7200	7200	7200	7200

5.7.5   Incentive Mechanism

From the SE learning results presented in Sect. 5.7.2 we can see that, when different users have similar expectations, the recommendation quality solely can motivate users to provide enough ratings to the recommendation server, so that the server can generate satisfying recommendations for all users. In this case, external incentive for user participation is not necessary. However, if there are significant differences among users' expectations and users no longer rate more items after they are satisfied, then the satisfaction equilibrium cannot be achieved via the proposed learning algorithm. In such case, some kind of external incentive (e.g. monetary rewards) is required to encourage users to provide more ratings.

As described in Sect. 5.7.1.1, the cost of choosing action $A^{(k)}$ is defined as $c_i\left(A^{(k)}\right) \triangleq \left|A^{(k)}\right|$. Suppose that the recommendation server pays $b\left(A^{(k)}\right) \triangleq \kappa \left|A^{(k)}\right|$ to the user as a reward, where the parameter $\kappa \in (0, 1)$ denotes the monetary reward that the user can get by rating one item. Paying rewards to users can be seen as a way to reduce the rating costs of users. More specifically, when user i gets a

Table 5.4 Simulation results of satisfaction equilibrium learning with rewards

	runID	1%: $\eta_i = 0.85$, 99%: $\eta_i = 0.5$					20%: $\eta_i = 0.85$, 80%: $\eta_i = 0.5$				
	runID	1	2	3	4	5	1	2	3	4	5
$\kappa = 0.01$	n_{stop}	102	172	184	201	102	401	302	303	304	304
	N_S	7200	7200	7200	7200	7200	7200	7200	7200	7200	7200
	$\bar{\sigma}_i$	0.939	0.888	0.891	0.935	0.938	0.96	0.964	0.973	0.98	0.979
$\kappa = 0.1$	n_{stop}	183	205	201	201	202	301	302	304	303	302
	N_S	7200	7200	7200	7200	7200	7200	7200	7200	7200	7200
	$\bar{\sigma}_i$	0.859	0.931	0.885	0.886	0.899	0.904	0.914	0.931	0.923	0.914
$\kappa = 0.5$	n_{stop}	15	19	16	16	18	21	18	20	20	21
	N_S	7200	7200	7200	7200	7200	7200	7200	7200	7200	7200
	$\bar{\sigma}_i$	0.884	0.907	0.889	0.89	0.904	0.916	0.903	0.911	0.912	0.915
$\kappa = 0.9$	n_{stop}	3	3	3	3	3	4	4	4	5	4
	N_S	7200	7200	7200	7200	7200	7200	7200	7200	7200	7200
	$\bar{\sigma}_i$	0.925	0.926	0.922	0.923	0.925	0.962	0.963	0.963	0.982	0.963

reward $b\left(A^{(k)}\right)$, the actual rating cost he pays is $c_i\left(A^{(k)}\right) - b\left(A^{(k)}\right)$. According to Algorithm 1, actions with low cost are preferred by users, hence users may rate more items if they are rewarded. Besides, motivated by the monetary rewards, users will continue to rate items even if they are satisfied with current recommendations. To formulate this intuition, we make a small modification to Algorithm 1: at each iteration, if the user is satisfied, i.e. $v_i\,(n-1) =1$, the probability that the user keeps previous action is defined as $\mu \triangleq 1 - \kappa$. Since $0 < \kappa < 1$, there is $\mu < 1$. As we have verified in Sect. 5.7.2, a satisfaction equilibrium can always be achieved when $\mu < 1$.

To evaluate the performance of the modified learning algorithm, we conduct simulations on Jester data set. Parameters of the algorithm are set in the same way as we've done before, and the reward parameter κ is set to 0.01, 0.1, 0.5, 0.9 respectively. Again, to reduce the influence of randomness, we run the algorithm five times for each setting. Simulation results are shown in Table 5.4. By comparing Tables 5.1 and 5.4 we can see that, when the recommendation server pays rewards to users, the learning algorithm converges to the equilibrium at a fast speed. The higher the rewards are, the faster the algorithm converges. For example, when there are only 1% of users who have high expectations for recommendation quality, if no reward is offered and $\mu = 0.9$, at least 400 iterations are required to reach an SE; if the recommendation server adopts a reward mechanism and sets $\kappa = 0.1$, the value of μ is still 0.9, but this time an SE can be achieved after about 200 iterations.

The simulation results indicate that the recommendation sever can push the interactions among users towards satisfaction equilibrium by offering rewards to users. The reward mechanism proposed above is quite simple. A more elaborate incentive mechanism, where the differences in rating cost and expectation for recommendation quality among users are considered, should be developed. We will investigate this problem in future work.

5.8   Conclusion

In this chapter we formulated the interaction among users in a CF system as a game in satisfaction form. To learn the satisfaction equilibrium of the game, we proposed a behavior rule that a user iteratively updates the probability distribution over his action space and gradually rate more items. We have analyzed the convergence of the proposed rule under some simplifying assumptions. By conducting simulations on real-world data, we have demonstrated that when users have similar expectations for the recommendation quality, a satisfaction equilibrium can be achieved via users' spontaneous rating behaviors.

The game-theoretic analysis we presented in this chapter may provide some implications to the study of user behaviors in collaborative systems. The derived convergence conditions may also be helpful to the design of incentive mechanisms. In future work, we'd like to investigate how to utilize both the intrinsic motivation and external incentives to guide users to share their data rationally.

References

1. J. Bobadilla, F. Ortega, A. Hernando, and A. Gutiérrez, "Recommender systems survey," *Knowledge-based systems*, vol. 46, pp. 109–132, 2013.
2. X. Su and T. M. Khoshgoftaar, "A survey of collaborative filtering techniques," *Advances in artificial intelligence*, vol. 2009, p. 4, 2009.
3. M. Grčar, D. Mladenič, B. Fortuna, and M. Grobelnik, "Data sparsity issues in the collaborative filtering framework," in *Proceedings of the 7th International Conference on Knowledge Discovery on the Web: Advances in Web Mining and Web Usage Analysis*, ser. WebKDD'05. Berlin, Heidelberg: Springer-Verlag, 2006, pp. 58–76.
4. D. Rafailidis and P. Daras, "The tfc model: Tensor factorization and tag clustering for item recommendation in social tagging systems," *IEEE Transactions on Systems, Man, and Cybernetics: Systems*, vol. 43, no. 3, pp. 673–688, May 2013.
5. M. Mao, J. Lu, G. Zhang, and J. Zhang, "Multirelational social recommendations via multigraph ranking," *IEEE Transactions on Cybernetics*, vol. PP, no. 99, pp. 1–13, 2016.
6. Y. Ren, G. Li, J. Zhang, and W. Zhou, "Lazy collaborative filtering for data sets with missing values," *IEEE Transactions on Cybernetics*, vol. 43, no. 6, pp. 1822–1834, Dec 2013.
7. B. Li, X. Zhu, R. Li, and C. Zhang, "Rating knowledge sharing in cross-domain collaborative filtering," *IEEE Transactions on Cybernetics*, vol. 45, no. 5, pp. 1068–1082, May 2015.
8. P. Symeonidis, "Clusthosvd: Item recommendation by combining semantically enhanced tag clustering with tensor hosvd," *IEEE Transactions on Systems, Man, and Cybernetics: Systems*, vol. 46, no. 9, pp. 1240–1251, Sept 2016.
9. W. Wu, R. Ma, and J. Lui, "Distributed caching via rewarding: An incentive scheme design in p2p-vod systems," *Parallel and Distributed Systems, IEEE Transactions on*, vol. 25, no. 3, pp. 612–621, March 2014.
10. Y. Gao, Y. Chen, and K. J. R. Liu, "On cost-effective incentive mechanisms in microtask crowdsourcing," *IEEE Transactions on Computational Intelligence and AI in Games*, vol. 7, no. 1, pp. 3–15, March 2015.
11. Y.-H. Yang, Y. Chen, C. Jiang, C.-Y. Wang, and K. Liu, "Wireless access network selection game with negative network externality," *Wireless Communications, IEEE Transactions on*, vol. 12, no. 10, pp. 5048–5060, October 2013.

12. R. Gibbons, *A primer in game theory.* Harvester Wheatsheaf Hertfordshire, 1992.
13. L. Xu, C. Jiang, Y. Chen, Y. Ren, and K. J. R. Liu, "User participation game in collaborative filtering," in *2014 IEEE Global Conference on Signal and Information Processing (GlobalSIP)*, Dec 2014, pp. 263–267.
14. M. Halkidi and I. Koutsopoulos, "A game theoretic framework for data privacy preservation in recommender systems," in *Machine Learning and Knowledge Discovery in Databases.* Springer, 2011, pp. 629–644.
15. Y. Chen and K. Liu, "Understanding microeconomic behaviors in social networking: An engineering view," *Signal Processing Magazine, IEEE*, vol. 29, no. 2, pp. 53–64, March 2012.
16. E. Mojica-Nava, C. A. Macana, and N. Quijano, "Dynamic population games for optimal dispatch on hierarchical microgrid control," *IEEE Transactions on Systems, Man, and Cybernetics: Systems*, vol. 44, no. 3, pp. 306–317, March 2014.
17. S. M. Perlaza, H. Tembine, S. Lasaulce, and M. Debbah, "Quality-of-service provisioning in decentralized networks: A satisfaction equilibrium approach," *Selected Topics in Signal Processing, IEEE Journal of*, vol. 6, no. 2, pp. 104–116, 2012.
18. S. Ross and B. Chaib-draa, "Satisfaction equilibrium: Achieving cooperation in incomplete information games," in *Advances in Artificial Intelligence.* Springer, 2006, pp. 61–72.
19. K. Goldberg, T. Roeder, D. Gupta, and C. Perkins, "Eigentaste: A constant time collaborative filtering algorithm," *Information Retrieval*, vol. 4, no. 2, pp. 133–151, 2001.
20. R. M. Bell and Y. Koren, "Improved neighborhood-based collaborative filtering," in *KDD Cup and Workshop at the 13th ACM SIGKDD International Conference on Knowledge Discovery and Data Mining.* sn, 2007.
21. B. Sarwar, G. Karypis, J. Konstan, and J. Riedl, "Item-based collaborative filtering recommendation algorithms," in *Proceedings of the 10th International Conference on World Wide Web*, ser. WWW '01. New York, NY, USA: ACM, 2001, pp. 285–295. [Online]. Available: http://doi.acm.org/10.1145/371920.372071
22. Y. Koren, R. Bell, and C. Volinsky, "Matrix factorization techniques for recommender systems," *Computer*, vol. 42, no. 8, pp. 30–37, Aug 2009.
23. T. Kandappu, A. Friedman, R. Boreli, and V. Sivaraman, "Privacycanary: Privacy-aware recommenders with adaptive input obfuscation," in *2014 IEEE 22nd International Symposium on Modelling, Analysis Simulation of Computer and Telecommunication Systems*, Sept 2014, pp. 453–462.
24. N. Polatidis, C. K. Georgiadis, E. Pimenidis, and H. Mouratidis, "Privacy-preserving collaborative recommendations based on random perturbations," *Expert Systems with Applications*, vol. 71, pp. 18–25, 2017. [Online]. Available: http://www.sciencedirect.com/science/article/pii/S0957417416306479
25. J. Parraarnau, D. Rebollomonedero, and J. Forné, "Optimal forgery and suppression of ratings for privacy enhancement in recommendation systems," *Entropy*, vol. 16, no. 3, pp. 1586–1631, 2014.
26. P. De Meo, G. Quattrone, G. Terracina, and D. Ursino, "An xml-based multiagent system for supporting online recruitment services," *Trans. Sys. Man Cyber. Part A*, vol. 37, no. 4, pp. 464–480, Jul. 2007. [Online]. Available: http://dx.doi.org/10.1109/TSMCA.2007.897696
27. H. Blanco and F. Ricci, "Acquiring user profiles from implicit feedback in a conversational recommender system," in *Proceedings of the 7th ACM Conference on Recommender Systems*, ser. RecSys '13. New York, NY, USA: ACM, 2013, pp. 307–310. [Online]. Available: http://doi.acm.org/10.1145/2507157.2507217

Chapter 6
Privacy-Accuracy Trade-Off in Distributed Data Mining

Abstract An important issue in distributed data mining is privacy. It is necessary for each participant to make sure that its privacy is not disclosed to other participants or a third party. To protect privacy, one can apply a differential privacy approach to perturb the data before sharing them with others, which generally hurts the mining result. That is to say, the participant faces a trade-off between privacy and the mining result. In this chapter, we study a distributed classification scenario where a mediator builds a classifier based on the perturbed query results returned by a number of users. A game theoretical approach is proposed to analyze how users choose their privacy budgets. Specifically, interactions among users are modeled as a game in satisfaction form. And an algorithm is proposed for users to learn the satisfaction equilibrium (SE) of the game. Experimental results demonstrate that, when the differences among users' expectations are not significant, the proposed learning algorithm can converge to an SE, at which every user achieves a balance between the accuracy of the classifier and the preserved privacy.

6.1 Introduction

With the development of Internet and cloud computing, distributed data mining, which extracts knowledge from distributed data sources [1], becomes more common in recent years. By sharing data with others and conducting mining on a joint data set, users who participate in distributed data mining can get more useful knowledge than that they can get from their own data.

However, distributed data mining can lead to serious privacy problems if the data contain sensitive information of the participants. For example, in order to get a better understanding of customers' purchasing behavior, several retailers may conduct data mining on the collective of customer data. If the retailers are competitors in the market, then no one wants to disclose much information about its own customers to others. How to perform distributed data mining whilst preserving the privacy of participants is an important issue.

To deal with the privacy issues in data mining, researchers have proposed various methods to realize privacy-preserving data mining (PPDM) [2, 3]. Techniques such as data perturbation [4] and encryption [5] are often applied to PPDM. Specifically,

© Springer International Publishing AG, part of Springer Nature 2018   151
L. Xu et al., *Data Privacy Games*, https://doi.org/10.1007/978-3-319-77965-2_6

as pointed out in [6], the problem of privacy-preserving distributed data mining is closely related to a subfield of cryptography named *secure multi-party computation* (SMC) [7]. In SMC, a number of participants, each of which has a private data, want to compute the value of a function which takes all the private data as input. A secure protocol is established to control the information exchange between the participants and make sure that every participant only knows its own data and the computation result. SMC protocols are widely applied in the study of privacy-preserving distributed data mining [8–10].

An implicit assumption of SMC is that there is no trusted third party that can do the computation for all participants. The success of SMC depends on whether the participants behave honestly. That is, if some participant deviates from the secure protocol or tries to learn extra information from the information received from others, other participants' privacy will be compromised. How to design a secure protocol that is robust to participants' dishonest behaviors is a complicated problem. In this chapter, we consider a distributed data mining scenario where a mediator assists multiple users to conduct the mining task. Here we take classification as an example. The mediator builds a classifier based on the data provided by users, and each user only communicates with the mediator. Since there is no information change between users, it is hard for a user to probe others' privacy, and we can only focus on how to prevent the privacy disclosure incurred by the communication between the user and the mediator.

Suppose that the mediator is untrustworthy, in the sense that it may try to learn users' sensitive information from users' data. In such a case, the user needs to adopt some measure to prevent the mediator from learning its privacy. In current study of data privacy, differential privacy [11, 12] has become the de facto standard of privacy definition. Following the principal of ε-differential privacy, we propose a distributed Naïve Bayes classification algorithm. To build a classifier, the mediator sends count queries to each user. The user runs queries on its own data and adds noise, which is determined by the *privacy budget* ε, to the results. By adding noise, the user can protect its privacy to some extent. While this will hurt the accuracy of the classifier. It is necessary for the user to make a trade-off between classification accuracy and privacy security. On the other hand, the accuracy of the classifier depends on all the query results provided by users. It is possible that the user cannot get a satisfying accuracy even if it provides true query results to the mediator, since other users may have added large noise to their results. Hence, when a user makes decisions on the privacy budget, it should take other users' decisions into account. The users are actually interacting with each other through the mediator. Also, users are usually rational, in the sense that every user wishes to obtain high classification accuracy without revealing much private information of its own. Therefore, we can employ game theory [13] to model the distributed classification scenario.

In this chapter, we propose a game model to analyze users' behaviors in distributed differentially-private classification [14]. Different from previous game theoretical approaches which usually establish a game model with complete information, here we model the interactions among users as a game with incomplete information. Specifically, it is assumed that each user only has knowledge of its own data and the accuracy of the classifier, while how other users' choose their privacy budgets cannot be observed. To analyze the equilibrium of this game, we adopt the notion of *satisfaction equilibrium* (SE) that was originally proposed by Ross and Chaib-draa [15]. A game is said to be in SE when all players simultaneously satisfy their individual constrains. In the context of distributed classification, we treat a user's expectation for classification accuracy as its constrain. Inspired by previous studies [16, 17], we propose a learning algorithm for users that can lead to an SE of the proposed game. Simulation results on real data show that the SE learning can help the user to make a privacy-preserving decision.

The rest of the chapter is organized as follows. Section 6.2 briefly introduces some studies that are related to our work. Section 6.3 describes the system model and Sect. 6.4 presents in details the game formulation. In Sect. 6.5, we present the algorithm proposed to learn the satisfaction equilibrium. The convergence analysis is conducted in Sect. 6.6. Simulation results are shown in Sect. 6.7. Finally, conclusions are drawn in Sect. 6.8.

6.2 Related Work

6.2.1 Game Theory

Game theory provides a formal approach to model the interactions among a group of agents who have to choose optimal actions considering the effects of other agents' decisions [13]. Researchers have applied game theory to the privacy problems in data mining. In [18], Kargupta et al. formalize the SMC problem as a static game with complete information. By analyzing the Nash equilibriums, they propose a cheap-talk based protocol that can prevent collusion among users. Miyaji et al. [19] propose a two-party secure set-intersection protocol in a game theoretic setting. In [20], Ge et al. propose a SMC-based algorithm for privacy-preserving distributed association rule mining (PPDARM) which employs a secret sharing technique to prevent collusion. The secret sharing scenario is modeled as a repeated game in Nanvati and Jinwala's work [21]. In [22], Xu et al. model the interaction between a data user and a data collector as a sequential game with complete and perfect information. And they applied backward induction to find the game's subgame perfect Nash equilibriums. Previous studies generally model the interactions among users as a game with complete information. While in this chapter, we propose a game with incomplete information. Each user can only observe the classification result, and there is no information exchange between users. Hence the user has no way to learn other users' strategies.

6.2.2 Mechanism Design

Mechanism design considers how to implement good system-wide solutions to problems that involve multiple self-interested agents with private information about their preferences for different outcomes [23]. Some researchers have applied mechanism design to privacy-preserving distributed data mining. In [24], Nix and Kantarcioglu propose two incentive compatible mechanisms to encourage users to share true data in distributed data mining. Based on Nix and Kantarcioglu's work, Panoui et al. [25] propose a Vickrey-Clarke-Groves (VCG) mechanism for privacy-preserving collaborative classification. In their model, a data provider can choose to provide true data, perturbed data, or randomized data. They show that the VCG mechanism can lead to high accuracy of the data mining task, meanwhile the privacy of data providers can be preserved. Similar to their work, we also consider that users can provide perturbed data so as to protect their privacy. However, instead of designing a mechanism towards high accuracy, we assume that different users have different expectations for the accuracy. As long as a satisfying result can be produced, the user prefers to provide few true data.

6.3 System Model

In this section, we describe the distributed classification model in details. Suppose that data are either horizontally distributed or vertically distributed. We first consider the former scenario. As shown in Fig. 6.1, a set of users $\mathcal{U} = \{u_1, u_2 \cdots, u_N\}$ interact with a mediator. Each user $u_l \in \mathcal{U}$ has a set of data records, denoted as D_l. Each data record consists of M attributes. Traditionally, users provide their data

Fig. 6.1 A typical distributed classification scenario. The mediator builds a classifier on the data provided by users, and returns the classifier together with the evaluation result to users

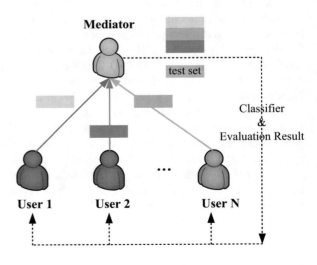

to the mediator, and the mediator trains a classifier on the collective data. However, considering that the sensitive information contained in the data may be disclosed, users generally refuse to share their data directly.

6.3.1 Naïve Bayes Classification

Suppose that the mediator applies the Naïve Bayes algorithm [26] to train the classifier. Based on the assumption that given the class, each attribute is conditionally independent of each of the other attributes, the Naïve Bayes algorithm applies the Bayes rule to predict the class of the data. Specifically, denote the M attributes as $X_1, X_2, \cdots, X_M$ respectively, and denote the class variable as Y. Suppose there are K classes. Given a new data record $\mathbf{X} \triangleq (x_1, x_2, \cdots, x_M)$ with x_i ($i = 1, \cdots, M$) denoting the value of the attribute X_i, the class of the record is predicted by

$$y = \underset{y_k}{\text{argmax}} \ \Pr\left(Y = y_k\right) \prod_{i=1}^{M} \Pr\left(X_i = x_i | Y = y_k\right). \tag{6.1}$$

Above equation indicates that the core of training a Naïve Bayes classifier is to estimate the following two types of distributions from the training data. The first type is the distribution of class. Given a training set D, for each $k \in \{1, 2, \cdots, K\}$, the prior probability that a data record belongs to the class y_k can be estimated by

$$\widehat{\Pr}\left(Y = y_k\right) = \frac{\#D\left\{Y = y_k\right\}}{|D|}, \tag{6.2}$$

where $|D|$ denotes the number of records in D, and the operator $\#D\{s\}$ returns the number of records in D that satisfy the property s.

The second type is the distribution of each attribute given a specific class. Here we assume that all the attributes take discrete values. Each attribute X_i has J_i possible values $x_{i1}, x_{i2}, \cdots, x_{iJ_i}$. Given a class y_k, the distribution of the attribute X_i can be estimated by

$$\widehat{\Pr}\left(X_i = x_{ij} | Y = y_k\right) = \frac{\#D\left\{X_i = x_{ij}, Y = y_k\right\}}{\#D\left\{Y = y_k\right\}}. \tag{6.3}$$

6.3.2 Differential Privacy

As a mathematically rigorous definition of privacy, differential privacy [11] is now widely applied in the study of data privacy. Roughly speaking, differential privacy guarantees that the removal or addition of a single record does not significantly affect the statistic results from the data set. A randomized function $\mathscr{K}$ is said to be ε-differentially private [12] if for all data sets D_1 and D_2 differing on at most one record, there is

$$\Pr(\mathscr{K}(D_1) \in S) \leq e^{\varepsilon} \Pr(\mathscr{K}(D_2) \in S) , \qquad (6.4)$$

where S denotes a subset of the range of function $\mathscr{K}$.

A count query, which counts the number of record in a data set satisfying a certain property, can be seen as a function f_Q mapping a data set to an integer number. By applying f_Q to a data set D, one can get a result $f_Q(D)$. To prevent the query results disclosing information about individual records, one can add appropriately chosen random noise to the real result. Specifically, by adding a number randomly drawn from a Laplace distribution $Lap\left(\frac{1}{\varepsilon}\right)$ to the result $f_Q(D)$, one can realize the ε-differential privacy on the query function f [12]. The parameter ε is usually referred to as the *privacy budget*. The smaller the budget is, the less accurate the result is, and the more privacy the result can preserve.

6.3.3 Differentially-Private Classification

From the above discussion we can see that, training a Naïve Bayes classifier is essentially running a series of count queries on the training set. This implies that as long as the mediator can get the statistics from users, it doesn't have to acquire the raw data. With this in mind, we propose the following method to train a classifier in a distributed and differentially-private manner.

The mediator, who has the knowledge of the set of classes $\{y_1, \cdots, y_K\}$, the set of attributes $\{X_1, \cdots, X_M\}$, and the possible values of each attribute $\{x_{i1}, \cdots, x_{iJ_i}\}$, defines two sets of count queries. The first set Q_C is about the class distributions. Each query $\#D\{Y = y_k\}$ in Q_C corresponds to a class. The second set Q_A is about the attribute distributions, and each query $\#D\{X_i = x_{ij}, Y = y_k\}$ corresponds to a specific value of a specific attribute and a specific class. For each user u_l, the mediator first sends all the queries in Q_C to the user. The user runs each query $\#D\{Y = y_k\}$ on its data D_l and directly returns the result α_{lk} to the mediator. Then the mediator sends the queries in Q_C to the user. Again, the user runs each query $\#D\{X_i = x_{ij}, Y = y_k\}$ on its data and gets the result β_{lijk}. But before sending the result to the mediator, the user adds Laplacian noise to the result, so as to meet the requirement of differential privacy. Specifically, the user u_l randomly draws a number σ_{lijk} based on the distribution $Lap\left(\frac{1}{\varepsilon_l}\right)$, where ε_l is the privacy budget chosen by u_l. Then the user returns $\tilde{\beta}_{lijk} \triangleq \beta_{lijk} + \sigma_{lijk}$ to the mediator. In addition to the results of queries, the user also sends the number of records in its data set, namely $|D_l|$, to the mediator.

After receiving the results from all users, the mediator can build a classifier by using (6.2) and (6.3), where $|D|$, $\#D\{Y = y_k\}$ and $\#D\{X_i = x_{ij}, Y = y_k\}$ are computed as

$$|D| = \sum_{l=1}^{N} |D_l| , \qquad (6.5)$$

$$\#D\{Y = y_k\} = \sum_{l=1}^{N} \alpha_{lk} , \qquad (6.6)$$

$$\#D\{X_i = x_{ij}, Y = y_k\} = \sum_{l=1}^{N} \tilde{\beta}_{lijk} . \qquad (6.7)$$

Suppose that the mediator has an independent data set that can be used to evaluate the performance of the classifier. Here we choose *accuracy* to indicate the performance. After the training and the evaluation, the mediator returns the data mining result, i.e. the classifier together with the accuracy $\gamma \in [0, 1]$, to all the users. A simple illustration of above procedure is shown in Fig. 6.2.

6.3.4 Vertically Distributed Data

Above we have discussed how to implement differentially private classification where data are horizontally distributed among multiple users. When data are

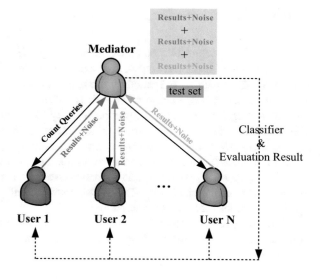

Fig. 6.2
Differentially-private distributed classification. The mediator sends count queries to all users, and builds a classifier based on the perturbed query results

vertically distributed, a similar classification approach can be applied. Suppose that each user in $\mathscr{U}$ has L data records. The data record owned by the user u_l is characterized by a set of M_l attributes, denoted as Atr_l. We assume that for any u_l and $u_{l'}$ ($l \neq l'$), there is $Atr_l \cap Atr_{l'} = \varnothing$. Hence the total number of attributes is

$$M \triangleq \sum_{l=1}^{N} M_l.$$

To train a Naïve Bayes classifier, the mediator sends count queries to each user. For the user u_l, the mediator first sends queries $\#D\{Y = y_k\}$ ($k = 1, \cdots, K$). Then for each attribute $X_i \in Atr_l$, the mediator sends a set of queries $\#D\{X_i = x_{ij}, Y = y_k\}$ to the user. Similar as before, the user runs each query $\#D\{Y = y_k\}$ on its data and directly returns the result α_{lk} to the mediator. While for the query $\#D\{X_i = x_{ij}, Y = y_k\}$, the user adds noise σ_{lijk}, which is determined by its privacy budget ε_l, to the real result.

After receiving the results of all queries from all users, the mediator applies (6.2) and (6.3) to build the classifier, where $|D| = L$, $\#D\{Y = y_k\}$ is computed by (6.6), and $\#D\{X_i = x_{ij}, Y = y_k\}$ is computed by (6.7). Then the mediator evaluates the accuracy of the classifier on the test set.

From the training procedures described above we can see that, the performance of the classifier heavily depends on the noise added to query results. Thus, whether a user can get a good classification result depends on not only its own choice of the privacy budget but also the privacy budgets chosen by other users. We can say that the users interact with each other via the mediator. Next we will apply game theory to formulate the interactions among users. The game formulation can be applied to both the horizontal distribution case and the vertical distribution case.

6.4 Satisfactory Game

6.4.1 Game Formulation

Player and *action* are two basic elements of a game [13]. In the aforementioned distributed classification scenario, all the users in $\mathscr{U}$ are *players*. The privacy budget ε_l chosen by a user u_l is treated as the user's *action*. As introduced in the above section, the budget ε_l determines the distribution of the noise added to the query result. A larger ε_l implies a higher probability of adding zero noise to the query result. That is to say, the accuracy of the classifier increases with ε_l. As we will see in Sect. 6.7, when ε_l exceeds some threshold ε_{max}, increasing ε_l no longer improves the classification accuracy significantly. Therefore, we define the range of ε_l as $[\varepsilon_{min}, \varepsilon_{max}]$, where ε_{min} is a small constant. The range of ε_l, denoted as $\mathscr{A}_l$, is the *action space* of user u_l. The action space actually denotes the set of all the possible actions that the user can choose.

As described in (6.5)–(6.7), the mediator builds the Naïve Bayes classifier based on the query results returned by all users. And the query results depend on the privacy budget chosen by each user. Denote the *action profile* of all users, namely the combination of users' actions, as $\boldsymbol{\varepsilon} = (\varepsilon_1, \varepsilon_2, \cdots, \varepsilon_N)$. Then, given a test set, the accuracy of the classifier is determined by $\boldsymbol{\varepsilon}$. Considering this, we introduce a mapping $f : \mathscr{A} \rightarrow [0, 1]$ to represent the influence of users' actions on the accuracy of the classifier, where $\mathscr{A} = \mathscr{A}_1 \times \mathscr{A}_2 \times \cdots \times \mathscr{A}_N$. Then the accuracy can be written as

$$\gamma = f(\boldsymbol{\varepsilon}) = f(\varepsilon_l, \boldsymbol{\varepsilon}_{-l}), \tag{6.8}$$

where $\boldsymbol{\varepsilon}_{-l} = (\varepsilon_1, \cdots, \varepsilon_{l-1}, \varepsilon_{l+1}, \cdots, \varepsilon_N)$.

When every user chooses to use the largest privacy budget, i.e. to add the minimal noise to the query result, the resulting classifier can reach the best performance. Let $\boldsymbol{\varepsilon}^* \triangleq (\varepsilon_1^*, \cdots, \varepsilon_N^*)$ denote the special action profile with $\varepsilon_l^* = \varepsilon_{max}$ ($l = 1, 2, \cdots, N$), and $\gamma_{\max}$ denote the corresponding accuracy, i.e. $\gamma_{\max} = f(\boldsymbol{\varepsilon}^*)$. As long as there is one user who chooses a smaller privacy budget, the resulting accuracy is lower than $\gamma_{\max}$. Given other users' choices of privacy budget, the smaller ε_l is, the less accurate the query results $\#D\{Y = y_k\}$ and $\#D\{X_i = x_{ij}, Y = y_k\}$ will be, and the lower the classification accuracy will be. Similarly, given user u_l's choice ε_l, the accuracy of the classifier will decrease as the privacy budgets chosen by other users decrease. If we use a function $g(\cdot)$ to denote the relationship between privacy budget and the accuracy, i.e.

$$\gamma = g(\varepsilon_l, \varepsilon_{-l}), \tag{6.9}$$

where $\varepsilon_{-l} = \frac{1}{N-1} \sum_{j \neq l} \varepsilon_j$, then the above intuition can be described by the following assumption: for any user u_l, there is $\frac{\partial g(\varepsilon_l, \varepsilon_{-l})}{\partial \varepsilon_l} > 0$ and $\frac{\partial g(\varepsilon_l, \varepsilon_{-l})}{\partial \varepsilon_{-l}} > 0$.

6.4.2 Satisfaction Form

Due to the privacy concerns, users are generally reluctant to provide the real query results to the mediator. As a result, the best classification accuracy $\gamma_{\max}$ can rarely be realized. Different from previous studies [24, 25] where the participants of distributed data miming are assumed to pursue the best result, here we assume that the users just look forward to a satisfying result. That is, each user has an expectation, denoted as γ_l, for the accuracy of the classifier, and there is $\gamma_l < \gamma_{\max}$. So long as the realized γ is higher than γ_l, user u_l will be satisfied. Note that γ is determine by both the user u_l's action ε_l and other users' actions $\boldsymbol{\varepsilon}_{-l}$. Given $\boldsymbol{\varepsilon}_{-l}$, user u_l may get a satisfying result by choosing some action. Let $h_l(\boldsymbol{\varepsilon}_{-l})$ denote the set of such actions, i.e.

$$h_l(\boldsymbol{\varepsilon}_{-l}) = \{\varepsilon_l \in A_l : f(\varepsilon_l, \boldsymbol{\varepsilon}_{-l}) \geq \gamma_l\}. \tag{6.10}$$

It should be pointed out that $h_l\,(\boldsymbol{\varepsilon}_{-l})$ may be an empty set in some cases. Consider the following scenario: all users in $\mathscr{U}$, except user u_l, decide to add the maximal noise to the query results, i.e. to choose the smallest privacy budget. Then no matter what privacy budget user u_l chooses, the accuracy of the resulting classifier may be lower than user u_l's expectation.

Based on above discussions, we can use the flowing triple to describe the game among the participants of the distributed classification:

$$G = (\mathscr{U}, \{A_l\}, \{h_l\})\,. \qquad (6.11)$$

Such a formulation of game is called *satisfaction* form. Based on Ross and Chaibdraa's work [15, 27], Perlaza et al. first formally introduced this special game formulation in [16]. So far, the satisfaction form has been applied in the study of wireless communication [28, 29] and collaborative filtering [30].

6.4.3   Satisfaction Equilibrium

An important concept in game theory is equilibrium. For a game in satisfaction form, the corresponding notion of equilibrium is the *satisfaction equilibrium* (SE):

Definition 1 (Satisfaction Equilibrium) An action profile $\boldsymbol{\varepsilon}^+$ is an equilibrium of the game $G = (\mathscr{U}, \{\mathscr{A}_l\}, \{h_l\})$, if $\forall u_l \in \mathscr{U}$, there is $\boldsymbol{\varepsilon}_l^+ \in h_l\left(\boldsymbol{\varepsilon}_{-l}^+\right)$.

When an SE is achieved, all users are satisfied and no one will change its action. Since it is assumed that every user's expectation is lower than the best result $\gamma_{\max}$, the action profile $\boldsymbol{\varepsilon}^*$ mentioned before is an SE of the proposed game. However, according to the definition of $\boldsymbol{\varepsilon}^*$, when $\boldsymbol{\varepsilon} = \boldsymbol{\varepsilon}^*$, users may suffer a great loss of privacy. Hence, $\boldsymbol{\varepsilon}^*$ is not an ideal outcome of the game. To achieve the balance between the security of privacy and the performance of the classifier, we need to find other instantiations of SE which can preserve more privacy.

6.5   Learning Satisfaction Equilibrium

The game described in above section is a game with incomplete information, in the sense that each player has no knowledge of other players' actions. Therefore, different from general equilibrium concepts in the context of complete information games, the satisfaction equilibrium arises as the result of a learning process, rather than the result of rational thinking on players' beliefs and observations [15]. In this section, we study how to achieve the satisfaction equilibrium via learning. Specifically, we propose a learning algorithm that converges towards satisfaction equilibria.

The learning algorithm basically describes an iterative process of information exchange between users and the mediator. Let $\varepsilon_l(n)$ denote the action that user u_l chooses at iteration n $(n = 0, 1, \cdots)$. Considering that users prefer to preserve as much privacy as possible, at the beginning of the learning process, all users choose the smallest privacy budget. That is, for each user u_l, the initial action $\varepsilon_l(0) = \varepsilon_{min}$. The initial action profile $\boldsymbol{\varepsilon}(0) \triangleq (\varepsilon_1(0), \cdots, \varepsilon_N(0))$ determines the initial accuracy $\gamma(0)$ of the classifier.

At each iteration n $(n \geq 1)$, the user u_l first checks current accuracy of the classifier, namely $\gamma(n-1)$. How the user chooses its next action $\varepsilon_l(n)$ depends on whether the accuracy meets the user's expectation γ_l. For ease of description, we introduce a binary variable $s_l(n-1)$ which is defined as

$$s_l(n-1) = \begin{cases} 1, & if \ \gamma(n-1) \geq \gamma_l, \\ 0, & otherwise. \end{cases} \tag{6.12}$$

If $s_l(n-1) = 0$, i.e. the user is unsatisfied with current result, then it is more reasonable for the user to choose a new action than to stick with the previous action $\varepsilon_l(n-1)$. Intuitively, the user should add less noise to the query results. More specifically, the newly chosen privacy budget $\varepsilon_l(n)$ should be larger than $\varepsilon_l(n-1)$. Meanwhile, the user prefers to choose a privacy budget which is slightly different from the previous one, so that the user can still preserve much privacy. Nevertheless, it is possible that the user chooses the previous action, if the user believes that it has made enough contribution to the community and the unsatisfying result is attributed to other users' actions. Especially, if the user already provides nearly accurate query results to the mediator, i.e. $\varepsilon_l(n-1) = \varepsilon_{max}$, the user will not change its action. Based on above discussions, we use the following rule to determine the action $\varepsilon_l(n)$ for user u_l: define the increment of privacy budget as

$$\Delta\varepsilon_l(n) = \varepsilon_l(n) - \varepsilon_l(n-1) . \tag{6.13}$$

The user u_l randomly chooses $\Delta\varepsilon_l(n)$ from $[0, \varepsilon_{max} - \varepsilon_l(n-1)]$ according to the following the distribution

$$\Pr(\Delta\varepsilon_l(n) \leq \delta) = - \frac{1}{[\varepsilon_{max} - \varepsilon_l(n-1)]^2} \delta^2$$
$$+ \frac{2}{[\varepsilon_{max} - \varepsilon_l(n-1)]} \delta , \tag{6.14}$$
$$\delta \in [0, \varepsilon_{max} - \varepsilon_l(n-1)] .$$

The probability density function corresponding to above cumulative distribution is given by

$$f(\delta) = -\frac{2}{[\varepsilon_{\max} - \varepsilon_l(n-1)]^2}\delta$$
$$+ \frac{2}{[\varepsilon_{\max} - \varepsilon_l(n-1)]},$$
$$\delta \in [0, \varepsilon_{\max} - \varepsilon_l(n-1)].$$
(6.15)

Above equation implies that the larger $\Delta\varepsilon_l(n)$ is, the lower the probability of $\Delta\varepsilon_l(n)$ being chosen by the user is. Once the increment $\Delta\varepsilon_l$ is chosen, the user's action is $\varepsilon_l(n)$ determined.

If $s_l(n-1) = 1$, i.e. the user is satisfied with current result, then there is no need for the user to change its action. Actually, the user can just opt out of the system. But still, we consider the possibility that the user would like to reduce the noise added to query results so that other users and itself can be benefited. With this in mind, we define the following update rule of $\varepsilon_l(n)$. Similar as before, the user first randomly draws a value for $\Delta\varepsilon_l(n)$ from $[0, \varepsilon_{\max} - \varepsilon_l(n-1)]$. Different from the previous case where $s_l(n-1) = 0$, now there is a high possibility that $\varepsilon_l(n) = \varepsilon_l(n-1)$, since the user is already satisfied. Hence, we define

$$Pr(\Delta\varepsilon_l(n) = 0) = \kappa$$
(6.16)

and

$$Pr(0 < \Delta\varepsilon_l(n) \leq \delta) = -\frac{1-\kappa}{[\varepsilon_{\max} - \varepsilon_l(n-1)]^2}\delta^2$$
$$+ \frac{2(1-\kappa)}{[\varepsilon_{\max} - \varepsilon_l(n-1)]}\delta,$$
$$\delta \in (0, \varepsilon_{\max} - \varepsilon_l(n-1)].$$
(6.17)

The parameter κ in above two equations denotes to what extent a satisfied user would keep its previous action, and we define $0.5 < \kappa \leq 1$. After the value of $\Delta\varepsilon_l$ is determined, the user can choose the action $\varepsilon_l(n)$ correspondingly.

Above we have discussed how the user u_l chooses its action $\varepsilon_l(n)$ in different cases. After every user chooses its action, the mediator re-sends the count queries to all users and trains the classifier by using the returned results. Then the classifier is evaluated on the test set. After that, the updated classifier and the corresponding accuracy $\gamma_l(n)$ is published to all users.

A summary of above learning process is presented in Algorithm 1. Suppose that after a number of iterations, all users are satisfied with the mining result, then the iterative process stops. Let n_s denote the number of iterations. We say the learning algorithm converges to a SE $\boldsymbol{\varepsilon}^+ = (\varepsilon_1(n_s), \cdots, \varepsilon_N(n_s))$.

Algorithm 1 Learning the SE of the Game $G = (\mathscr{U}, \{\mathscr{A}_l\}, \{h_l\})$

1: $n = 0$;
2: $\varepsilon_l(0) = \varepsilon_{min}$;
3: **for all** $n > 0$ **do**
4:   compute the distribution of $\Delta \varepsilon_l$:
$\delta_{\max} = \varepsilon_{\max} - \varepsilon_l(n-1)$,
if $s_l(n-1) = 0$,

$$\Pr(\Delta \varepsilon_l(n) \leq \delta) = -\frac{1}{\delta_{\max}^2}\delta^2 + \frac{2}{\delta_{\max}}\delta \,, \delta \in [0, \delta_{\max}]$$

else

$$\Pr(\Delta \varepsilon_l(n) = 0) = \kappa,$$

$$\Pr(0 < \Delta \varepsilon_l(n) \leq \delta) = -\frac{1-\kappa}{\delta_{\max}^2}\delta^2 + \frac{2(1-\kappa)}{\delta_{\max}}\delta \,, \delta \in [0, \delta_{\max}]$$

5:   draw $\Delta \varepsilon_l(n)$ from $[0, \varepsilon_{\max} - \varepsilon_l(n-1)]$;
6:   $\varepsilon_l(n) = \varepsilon_l(n-1) + \Delta \varepsilon_l(n)$;
7: **end for**

6.6 Convergence of the Learning Algorithm

In this section, we present a simple analysis of the convergence of the proposed learning algorithm.

At the beginning of iteration n, the accuracy of the classifier is $\gamma(n-1)$. Based on users' expectations on the accuracy, users can be divided into two sets: the set of satisfied users, denoted as $\mathscr{N}_S(n) \triangleq \{u_l | u_l \in \mathscr{U}, \gamma(n-1) \geq \gamma_l\}$, and the set of unsatisfied users, denoted as $\mathscr{N}_U(n) \triangleq \{u_l | u_l \in \mathscr{U}, \gamma(n-1) < \gamma_l\}$. Consider the user u_i who has the highest expectations among all users. Suppose that at iteration n, all users except u_i are satisfied, i.e. $\mathscr{N}_U(n) = \{u_i\}$. Let u_j denote the user who has the highest expectation in $\mathscr{N}_S(n)$, i.e. $\gamma_j = \max_{u_k \in \mathscr{N}_S(n)} \gamma_k$. Then there is $\gamma_j \leq \gamma(n-1) < \gamma_i$. According to (6.9), there is

$$\gamma(n-1) = g(\varepsilon_i(n-1), \varepsilon_{-i}(n-1)), \tag{6.18}$$

where

$$\varepsilon_{-i}(n-1) = \frac{1}{N-1}\sum_{k \neq i} \varepsilon_k(n-1). \tag{6.19}$$

As mentioned before, the function $g(\cdot, \cdot)$ increases with ε_i and ε_{-i}. Thus, whether the algorithm can achieve a SE depends on how ε_i and ε_{-i} change during the iterative process.

According to Algorithm 1, when $\kappa < 1$, it is possible that the user in $\mathcal{N}_S(n)$ returns a more accurate result in response to the mediator's query in subsequent iterations, which means $\varepsilon_{-i}(n-1)$ will increase with n. In the meantime, the unsatisfied user will increase its privacy budget, hence $\varepsilon_i(n-1)$ also increases with n. As a result, after a number of iterations, the accuracy will meet user u_i's expectation, and a SE is achieved.

When $\kappa = 1$, users in $\mathcal{N}_S(n)$ no longer change their privacy budgets, which means $\forall n' \geq n, \varepsilon_{-i}(n'-1) = \varepsilon_{-i}(n-1)$. Therefore, whether the unsatisfied user u_i can get a satisfied result completely depends on the user itself. According to Algorithm 1, the user will gradually increase its privacy budget until it is satisfied. Suppose that at iteration m $(m > n)$, there is $\varepsilon_i(m-1) = \varepsilon_{max}$, i.e. the user has already chosen the maximal privacy budget. The corresponding accuracy is

$$
\begin{aligned}
\gamma(m-1) &= g(\varepsilon_i(m-1), \varepsilon_{-i}(m-1)) \\
&= g(\varepsilon_{max}, \varepsilon_{-i}(n-1))
\end{aligned} \tag{6.20}
$$

According to the classification algorithm described in Sect. 6.3.3, the accuracy of the classifier is determined by the total noise added to each query. For example, (6.7) can be rewritten as

$$
\#D\{X_i = x_{ij}, Y = y_k\} = \sum_{l=1}^{N} \beta_{lijk} + \sum_{l=1}^{N} \sigma_{lijk}. \tag{6.21}
$$

Though each user chooses its privacy budget independently, from the perspective of the mediator, the total noise $\sigma_{ijk} \triangleq \sum_{l=1}^{N} \sigma_{lijk}$ is determined by a certain privacy budget ε_U. That is to say, for the mediator, σ_{ijk} is drawn from a distribution $Lap\left(\frac{1}{\varepsilon_U}\right)$. At iteration n, the classification accuracy $\gamma(n-1)$ can be written as

$$
\gamma(n-1) = r(\varepsilon_U(n-1)) = g(\varepsilon_i(n-1), \varepsilon_{-i}(n-1)), \tag{6.22}
$$

where $r(\cdot)$ is an increasing function of the privacy budget ε_U. Then we get

$$
\begin{aligned}
&\gamma(m-1) - \gamma(n-1) \\
&= r(\varepsilon_U(m-1)) - r(\varepsilon_U(n-1))
\end{aligned} \tag{6.23}
$$

Since iteration n, only the user u_i has changed the noise added to query results. Hence, the difference between $\gamma(m-1)$ and $\gamma(n-1)$ is actually determined by $\varepsilon_{max} - \varepsilon_i(n-1)$. Denote the inverse function of $r(\cdot)$ as $r^{-1}(\cdot)$. Then we get

$$
\varepsilon_{max} - \varepsilon_i(n-1) = r^{-1}(\gamma(m-1)) - r^{-1}(\gamma(n-1)). \tag{6.24}
$$

If the algorithm converges at iteration m, i.e. $\gamma\,(m-1) \geq \gamma_i$, then there is

$$\varepsilon_i\,(n-1) \leq \varepsilon_{\max} - r^{-1}\,(\gamma_i) + r^{-1}\,(\gamma\,(n-1))\,. \qquad (6.25)$$

Above inequality implies that $\varepsilon_i\,(n-1)$ should be no bigger than the minimum of $\varepsilon_{\max} - r^{-1}\,(\gamma_i) + r^{-1}\,(\gamma\,(n-1))$. Considering that $\gamma\,(n-1) \geq \gamma_j$, we get

$$r^{-1}\,(\gamma_i) - r^{-1}\,(\gamma_j) \leq \varepsilon_{\max} - \varepsilon_i\,(n-1)\,. \qquad (6.26)$$

When the difference between γ_i and γ_j is too large, above inequality cannot hold, then the user u_i will never get a satisfying result. In other words, if there is a user whose expectation is significantly higher than other users' expectations, and satisfied users no longer change their privacy budgets, then the proposed learning algorithm cannot converge to an SE.

6.7 Simulation

To verify the convergence of the proposed SE learning algorithm, we have conducted a series of simulations on real-world data. In this section, we first describe the preparation of data, then we provide some experimental proofs for the assumption we've made for the learning algorithm. After that, we present a comparison of the learning results which are obtained under different settings of users' expectations.

6.7.1 Data Set

We choose three data sets, namely the adult data set, the car evaluation data set, and the handwritten digits data set, from UCI Machine Learning Repository [31] for simulation. The three data sets are widely used in the study of classification. In following descriptions, we refer to these data sets as *Adult*, *Car* and *Digits* respectively. Details of the data sets and corresponding settings of the classification experiments are given below.

6.7.1.1 Adult

The Adult data set was extracted from a census bureau database. After removing instances with unknown values, the data set is splitted into a training set, which consists of 30,162 instances, and a test set, which consists of 15,060 instances. Each instance is characterized by 14 attributes, including 6 integer attributes and 8 categorical attributes. We keep the categorical attributes for simulation. The instances are categorized into two classes. To simulate the horizontally distributed classification scenario, the training set is randomly divided into $N = 5$ parts. Each part consists of about 6000 instances. To simulate the vertically distributed scenario, the attributes are randomly divided into $N = 3$ groups, each of which consists of two or three attributes. Instances in the training set are divided correspondingly.

6.7.1.2   Car

The car evaluation data set was derived from a hierarchical decision model. There are 1728 instances in the data set. Each instance is characterized by six attributes, and all the attributes are categorical. The instances can be categorized into four classes. To perform classification, we randomly choose 1210 instances for training, and the rest 518 instances are used to evaluate the accuracy of the classifier. To simulate the horizontally distributed classification scenario, the training set is randomly divided into $N = 3$ parts. Each part consists of about 400 instances. To simulate the vertically distributed scenario, the attributes are randomly divided into $N = 3$ groups, each of which consists of two attributes. Instances in the training set are divided correspondingly.

6.7.1.3   Digits

This data set contains 3823 training instances and 1797 test instances. Each instance is characterized by 64 attributes. All attributes are integers in the range 0–16. Instances are categorized into ten classes, each of which corresponds to a digit. Considering that each instance represents a bitmap of a digit, we only simulate the horizontally distributed classification scenario, where the training set is randomly divided into $N = 3$ parts.

6.7.2   Relationship Between Classification Accuracy and Privacy Budget

A fundamental assumption of our study is that as the privacy budget chosen by the user increases, the query result obtained by the mediator becomes more accurate, and the performance of the classifier becomes better. The assumption is quite intuitive. Nevertheless, we conduct experiments to verify the assumption.

Suppose that there is only one user, who has all the training instances, interacting with the mediator. Given a training set, we run the classification algorithm described in Sect. 6.3.3 multiple times in different settings of privacy budget. Specifically, Starting from a very small constant $\varepsilon_{min} = 2.22 \times 10^{-16}$, the privacy budget ε gradually increases to 5. Each time after training the classifier for a certain ε, we apply the classifier to the test data and record the classification accuracy. To reduce the influence of randomness, for each ε we repeat the training-testing procedure for five times, and the average of the accuracy is reported. Figure 6.3a–c shows the experiment results obtained on *Adult*, *Car* and *Digits* respectively. As we can see, the accuracy increases with the privacy budget. The results confirm our assumption.

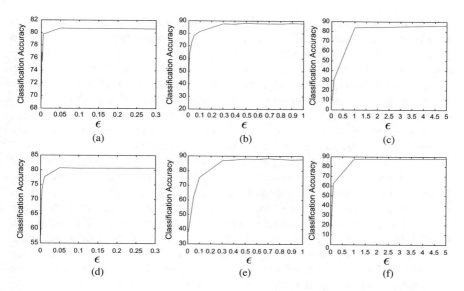

Fig. 6.3 The relationship between classification accuracy and privacy budget. (**a**) Adult ($N = 1$). (**b**) Car ($N = 1$). (**c**) Digits ($N = 1$). (**d**) Adult ($N = 5$). (**e**) Car ($N = 3$). (**f**) Digits ($N = 3$)

Considering that when simulating the equilibrium learning algorithm, we split the training data into multiple parts. To verify whether the assumption holds in the distributed case, we conduct another three groups of simulations on the three data sets respectively, where the training data are horizontally distributed among $N \in \{3, 5\}$ users. Given the training data, we set all users' privacy budgets to the same value ε, and run the classification algorithm described in Sect. 6.3.3 for five times to get an average result. Similar as before, ε gradually increases from ε_{min} to 5. From the results shown in Fig. 6.3d–f we can see that, in the distributed classification scenario, the classification accuracy also increases with the privacy budget. In addition, the simulation results show that after the privacy budget exceeds some threshold, there is no signification improvement in the classification accuracy. In subsequent experiments, the maximal privacy budget ε_{max} is set to 0.1 for Adult, 0.6 for Car, and 1 for Digits.

6.7.3 Users' Expectations

To verify the convergence of the proposed learning algorithm, we conduct simulations under different settings of users' expectations $\{\gamma_l\}_{l=1}^{N}$. Users' expectations are set in the following way. Given a data set and the number of users N, we first compute the best accuracy $\gamma_{\max}$ that can be achieved. That is, we use the original training set to train a classifier and then apply the classifier to the test set. Similar as before, we repeat above procedure for five times and record the average accuracy. Then, we compute the worst accuracy $\gamma_{\min}$ by setting the privacy budget to the maximum ε_{max}. After that, we divide the original training set into N parts with each part corresponding to a user. Users' expectations are randomly drawn from $[\gamma_{\min}, \gamma_{\max}]$. When $N = 3$, four groups of $\{\gamma_l\}_{l=1}^{N}$ are generated. The first group corresponds to the case where all users have low expectations. The second group corresponds to the case where all users have high expectations. The third and the fourth groups correspond to the cases where one user has a high expectation and the others have low expectations. When $N = 5$, in addition to the above four groups of users' expectations, one more group is generated, where two users have high expectations and the other three users have low expectations.

Before simulating the learning algorithm, we conduct the following experiment to see whether the learning process is necessary for achieving an equilibrium. Given the training data and a group of users' expectations $\{\gamma_l\}_{l=1}^{N}$, we let the users randomly choose privacy budgets from $[\varepsilon_{\min}, \varepsilon_{\max}]$. Then we evaluate the accuracy of the resulting classifier, and check if the accuracy satisfies all users' expectations. Denote the number of satisfied users as N_S. If $N_S = N$, then an SE is achieved. Above procedure is repeated multiple times. Due to the space limitation, here we only present the simulation results obtained under the following setting: data are horizontally distributed, and all users have high expectations. From the results shown in Fig. 6.4 we can see that, in many cases, there is a user who is unsatisfied with the classification accuracy. That is to say, if users just randomly choose their privacy budgets, the possibility of achieving an SE is low. While later we can see that, when the users behave according to the proposed learning algorithm, the high expectations of all users can be satisfied at some point.

6.7.4   Simulation Results of Equilibrium Learning

Given the training sets and a group of users' expectations, we run the learning algorithm under the setting of $\kappa = 0.9$ and $\kappa = 1$ respectively. The former simulates the situation that a satisfied user may increase its privacy budget, and the latter simulates the situation that a satisfied user no longer changes its privacy budget. Under each setting, we run the learning algorithm for five times to reduce the influence of randomness. In each run, the iterative process stops when all users are satisfied or the number of iterations reaches 500. After each run, we record the number of iterations, the classification accuracy and the average of users' privacy budgets.

Simulation results of the SE learning are shown in Figs. 6.5, 6.6, 6.7, 6.8 and 6.9, from which we can make the following observations:

6.7.4.1   When All the Users Have Similar Expectations for the Classification Accuracy

When all the users have similar expectations for the classification accuracy the learning algorithm can converge to an SE. Even when all users expect high

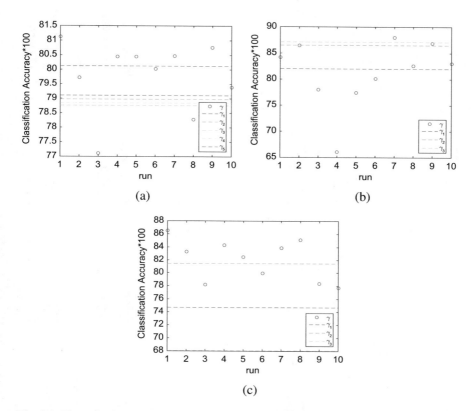

Fig. 6.4 The evaluation results of distributed classification with randomly chosen privacy budgets. The blue circle denotes the actual accuracy achieved by the classifier. The dotted lines denote users' expectations for the accuracy. (**a**) Adult ($N = 5$). (**b**) Car ($N = 3$). (**c**) Digits ($N = 3$)

accuracies and user becomes inactive after it is satisfied (i.e. $\kappa = 1$), an SE can still be achieved. For a given κ, when users' expectations become higher, generally it takes longer time for the algorithm to converge to the SE. And as expected, the average privacy budget becomes larger as the expectations become higher.

Given the setting of $\{\gamma_l\}_{l=1}^{N}$, by comparing the results of $\kappa = 0$ with those of $\kappa = 1$ we can see that, the values of average privacy budget are similar in the two cases. For example, according to Fig. 6.5e, when all the users have high expectations, the average privacy budget is about 0.031. The reason that the value of κ makes no difference to the learning results is that the differences among users' expectations are small. Before an SE is reached, all users are in an "unsatisfied" state, which means the users only apply (6.14) to determine the increment of privacy budget. With the increase of the accuracy, the users simultaneously become satisfied at some point. Hence the parameter κ has no influence to users' choices.

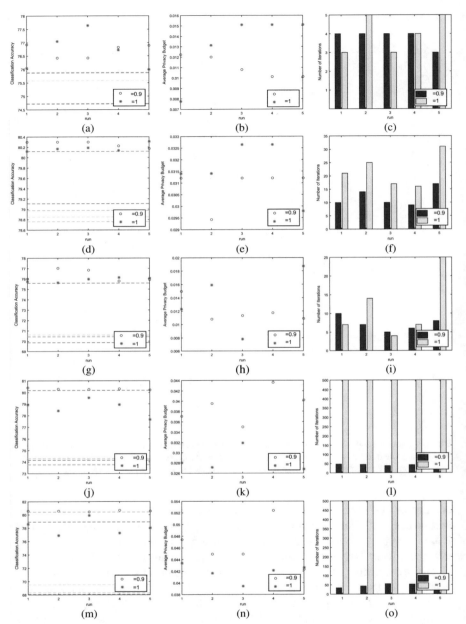

Fig. 6.5 Simulation results of satisfaction equilibrium learning on Adult. Data are horizontally distributed among five users. (**a**) Classification Accuracy. All users have low expectations. (**b**) Average Privacy Budget. All users have low expectations. (**c**) Number of Iterations. All users have low expectations. (**d**) Classification Accuracy. All users have high expectations. (**e**) Average Privacy Budget. All users have high expectations. (**f**) Number of Iterations. All users have high expectations. (**g**) Classification Accuracy. One user has relatively higher expectation. (**h**) Average Privacy Budget. One user has relatively higher expectation. (**i**) Number of Iterations. One user has relatively higher expectation. (**j**) Classification Accuracy. One user has a very high expectation. (**k**) Average Privacy Budget. One user has a very high expectation. (**l**) Number of Iterations. One user has a very high expectation. (**m**) Classification Accuracy. Two users have very high expectations. (**n**) Average Privacy Budget. Two users have very high expectations. (**o**) Number of Iterations. Two users have very high expectations

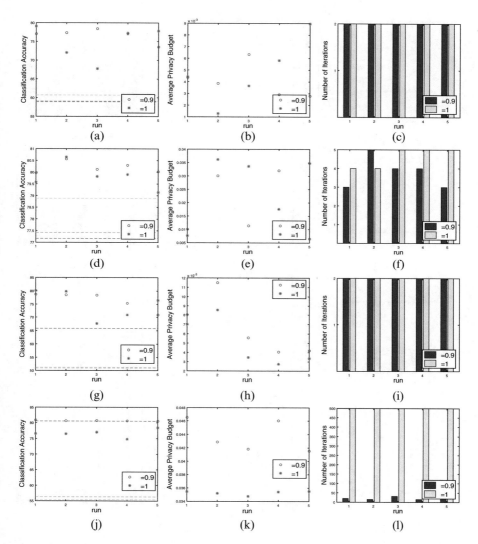

Fig. 6.6 Simulation results of satisfaction equilibrium learning on Adult. Data are vertically distributed among three users: (**a**) Classification Accuracy. All users have low expectations; (**b**) Average Privacy Budget. All users have low expectations; (**c**) Number of Iterations. All users have low expectations; (**d**) Classification Accuracy. All users have high expectations; (**e**) Average Privacy Budget. All users have high expectations; (**f**) Number of Iterations. All users have high expectations; (**g**) Classification Accuracy. One user has relatively higher expectation; (**h**) Average Privacy Budget. One user has relatively higher expectation; (**i**) Number of Iterations. One user has relatively higher expectation; (**j**) Classification Accuracy. One user has a very high expectation; (**k**) Average Privacy Budget. One user has a very high expectation; (**l**) Number of Iterations. One user has a very high expectation

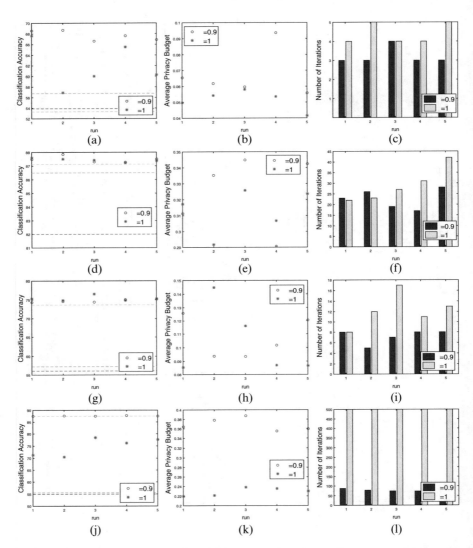

Fig. 6.7 Simulation results of satisfaction equilibrium learning on Car. Data are horizontally distributed among three users: (**a**) Classification Accuracy. All users have low expectations; (**b**) Average Privacy Budget. All users have low expectations; (**c**) Number of Iterations. All users have low expectations; (**d**) Classification Accuracy. All users have high expectations; (**e**) Average Privacy Budget. All users have high expectations; (**f**) Number of Iterations. All users have high expectations; (**g**) Classification Accuracy. One user has relatively higher expectation; (**h**) Average Privacy Budget. One user has relatively higher expectation; (**i**) Number of Iterations. One user has relatively higher expectation; (**j**) Classification Accuracy. One user has a very high expectation; (**k**) Average Privacy Budget. One user has a very high expectation; (**l**) Number of Iterations. One user has a very high expectation

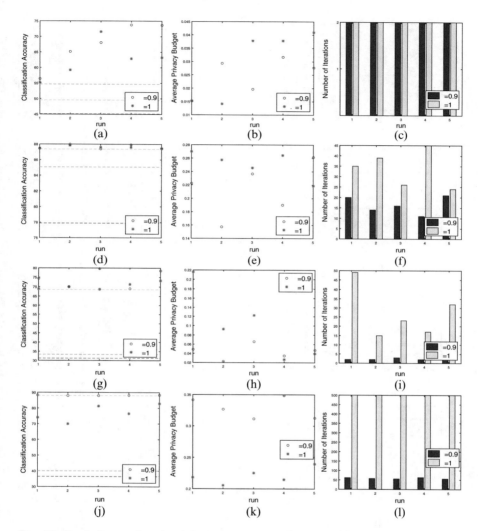

Fig. 6.8 Simulation results of satisfaction equilibrium learning on Car. Data are vertically distributed among three users: (**a**) Classification Accuracy. All users have low expectations; (**b**) Average Privacy Budget. All users have low expectations; (**c**) Number of Iterations. All users have low expectations; (**d**) Classification Accuracy. All users have high expectations; (**e**) Average Privacy Budget. All users have high expectations; (**f**) Number of Iterations. All users have high expectations; (**g**) Classification Accuracy. One user has relatively higher expectation; (**h**) Average Privacy Budget. One user has relatively higher expectation; (**i**) Number of Iterations. One user has relatively higher expectation; (**j**) Classification Accuracy. One user has a very high expectation; (**k**) Average Privacy Budget. One user has a very high expectation; (**l**) Number of Iterations. One user has a very high expectation

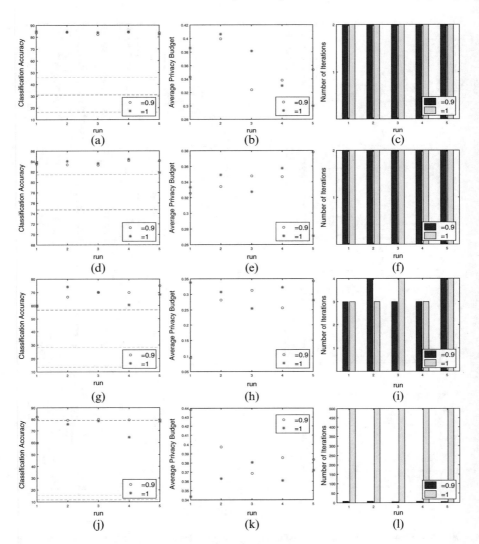

Fig. 6.9 Simulation results of satisfaction equilibrium learning on Digits. Data are horizontally distributed among three users: (**a**) Classification Accuracy. All users have low expectations; (**b**) Average Privacy Budget. All users have low expectations; (**c**) Number of Iterations. All users have low expectations; (**d**) Classification Accuracy. All users have high expectations; (**e**) Average Privacy Budget. All users have high expectations; (**f**) Number of Iterations. All users have high expectations; (**g**) Classification Accuracy. One user has relatively higher expectation; (**h**) Average Privacy Budget. One user has relatively higher expectation; (**i**) Number of Iterations. One user has relatively higher expectation; (**j**) Classification Accuracy. One user has a very high expectation; (**k**) Average Privacy Budget. One user has a very high expectation; (**l**) Number of Iterations. One user has a very high expectation

6.7.4.2   When Some User's Expectation Is Much Higher than Those of Others

When some user's expectation is much higher than those of others the algorithm may not converge to an SE. By comparing the results corresponding to the third and the fourth groups of expectations we can see that, when the difference between the high expectation and the low expectation is small, relative to the full range of the accuracy, an SE can be achieved, even if satisfied users no long increase their privacy budgets (i.e. $\kappa = 1$). While, when the difference between the expectations is significant, e.g. one expects an accuracy that approximates to the best value and the others' expectations are merely above the worst value, an SE cannot be achieved if $\kappa = 1$. Since the expectations of the $N - 1$ users are really low, these users become satisfied soon after the learning process starts. When they stops contributing to the classification, it is difficult for the rest user to make a big improvement in the accuracy. Similarly, as shown in Fig. 6.5m, when there two users who have very high expectations, the learning algorithm cannot converge to an SE either.

6.8   Conclusion

When participating in distributed data mining, users need to take measures to prevent privacy disclosure to other participants and the untrustworthy third party. In this chapter we formulated the interactions among users in a distributed classification scenario as a game in satisfaction form. To build a Naïve Bayes classifier, a mediator sends count queries to users. And users add noise to the results so as to meet the differential privacy criterion. When all users are satisfied with the accuracy of the classifier, a satisfaction equilibrium is achieved. To learn the equilibrium of the game, we proposed an algorithm which allows the user to iteratively change its privacy budget. By conducting simulations on real data sets, we have demonstrated that when users have similar expectations for the accuracy, the proposed learning algorithm can converge to a satisfaction equilibrium.

The game-theoretic analysis presented in this chapter may provide some implications to the design of incentive mechanisms which aim at encouraging users to provide accurate query results. We will investigate such incentive mechanisms in future work.

References

1. B.-H. Park and H. Kargupta, "Distributed data mining: Algorithms, systems, and applications," 2002, pp. 341–358.
2. L. Xu, C. Jiang, J. Wang, J. Yuan, and Y. Ren, "Information security in big data: Privacy and data mining," *IEEE Access*, vol. 2, pp. 1149–1176, 2014.

3. R. N., S. K., and A. Arul, "Survey on privacy preserving data mining techniques using recent algorithms," *International Journal of Computer Science and Information Technolo*, vol. 24, no. 9, pp. 1–7, 2017.
4. K. Liu, H. Kargupta, and J. Ryan, "Random projection-based multiplicative data perturbation for privacy preserving distributed data mining," *IEEE Transactions on Knowledge and Data Engineering*, vol. 18, no. 1, pp. 92–106, Jan 2006.
5. M. Kantarcioglu and C. Clifton, "Privacy-preserving distributed mining of association rules on horizontally partitioned data," *IEEE Transactions on Knowledge and Data Engineering*, vol. 16, no. 9, pp. 1026–1037, Sept 2004.
6. C. C. Aggarwal and P. S. Yu, "A general survey of privacy-preserving data mining models and algorithms," *Journal of Vascular Surgery*, vol. 8, no. 1, p. 6470, 2008.
7. R. Cramer, I. B. Damgrd, and J. B. Nielsen, *Secure Multiparty Computation and Secret Sharing*, 1st ed. New York, NY, USA: Cambridge University Press, 2015.
8. Y. Lindell and B. Pinkas, "Secure multiparty computation for privacy-preserving data mining," *Journal of Privacy and Confidentiality*, vol. 25, no. 2, pp. 761–766, 2009.
9. N. R. Nanavati and D. C. Jinwala, "A novel privacypreserving scheme for collaborative frequent itemset mining across vertically partitioned data," *Security and Communication Networks*, vol. 8, no. 18, pp. 4407–4420, 2015.
10. M. Sheikhalishahi and F. Martinelli, "Privacy-utility feature selection as a privacy mechanism in collaborative data classification," in *2017 IEEE 26th International Conference on Enabling Technologies: Infrastructure for Collaborative Enterprises (WETICE)*, June 2017, pp. 244–249.
11. C. Dwork, A. Roth *et al.*, "The algorithmic foundations of differential privacy," *Foundations and Trends® in Theoretical Computer Science*, vol. 9, no. 3–4, pp. 211–407, 2014.
12. C. Dwork, "Differential privacy: A survey of results," in *International Conference on Theory and Applications of Models of Computation*. Springer, 2008, pp. 1–19.
13. R. Gibbons, *A primer in game theory*. Harvester Wheatsheaf Hertfordshire, 1992.
14. L. Xu, C. Jiang, J. Li, Y. Zhao, and Y. Ren, "Privacy preserving distributed classification: A satisfaction equilibrium approach," in *IEEE GLOBECOM 2017*, Dec 2017, to appear.
15. S. Ross and B. Chaib-draa, "Satisfaction equilibrium: Achieving cooperation in incomplete information games," in *Advances in Artificial Intelligence*. Springer, 2006, pp. 61–72.
16. S. M. Perlaza, H. Tembine, S. Lasaulce, and M. Debbah, "Quality-of-service provisioning in decentralized networks: A satisfaction equilibrium approach," *Selected Topics in Signal Processing, IEEE Journal of*, vol. 6, no. 2, pp. 104–116, 2012.
17. J. Marden, H. Young, and L. Pao, "Achieving pareto optimality through distributed learning," in *Decision and Control (CDC), 2012 IEEE 51st Annual Conference on*, Dec 2012, pp. 7419–7424.
18. H. Kargupta, K. Das, and K. Liu, *Multi-party, Privacy-Preserving Distributed Data Mining Using a Game Theoretic Framework*. Berlin, Heidelberg: Springer Berlin Heidelberg, 2007, pp. 523–531.
19. A. Miyaji and M. S. Rahman, *Privacy-Preserving Data Mining: A Game-Theoretic Approach*. Berlin, Heidelberg: Springer Berlin Heidelberg, 2011, pp. 186–200.
20. X. Ge, L. Yan, J. Zhu, and W. Shi, "Privacy-preserving distributed association rule mining based on the secret sharing technique," in *Software Engineering and Data Mining (SEDM), 2010 2nd International Conference on*. IEEE, 2010, pp. 345–350.
21. N. R. Nanavati and D. C. Jinwala, "A novel privacy preserving game theoretic repeated rational secret sharing scheme for distributed data mining," *dcj*, vol. 91, p. 9426611777, 2013.
22. L. Xu, C. Jiang, J. Wang, Y. Ren, J. Yuan, and M. Guizani, "Game theoretic data privacy preservation: Equilibrium and pricing," in *2015 IEEE International Conference on Communications (ICC)*, June 2015, pp. 7071–7076.
23. D. C. Parkes, "Iterative combinatorial auctions: Achieving economic and computational efficiency," Ph.D. dissertation, University of Pennsylvania, 2001.

24. R. Nix and M. Kantarciouglu, "Incentive compatible privacy-preserving distributed classification," *IEEE Transactions on Dependable and Secure Computing*, vol. 9, no. 4, pp. 451–462, 2012.
25. A. Panoui, S. Lambotharan, and R. C.-W. Phan, "Vickrey-clarke-groves for privacy-preserving collaborative classification," in *Computer Science and Information Systems (FedCSIS), 2013 Federated Conference on*. IEEE, 2013, pp. 123–128.
26. T. M. Mitchell, *Machine Learning*, 1st ed. New York, NY, USA: McGraw-Hill, Inc., 1997.
27. S. Ross and B. Chaib-draa, "Learning to play a satisfaction equilibrium," in *Workshop on Evolutionary Models of Collaboration*, 2007.
28. R. Southwell, X. Chen, and J. Huang, "Quality of service games for spectrum sharing," *IEEE Journal on Selected Areas in Communications*, vol. 32, no. 3, pp. 589–600, March 2014.
29. Y. Sun, Y. Zhu, Z. Feng, and J. Yu, "Sensing processes participation game of smartphones in participatory sensing systems," in *2014 Eleventh Annual IEEE International Conference on Sensing, Communication, and Networking (SECON)*, June 2014, pp. 239–247.
30. L. Xu, C. Jiang, Y. Chen, Y. Ren, and K. J. R. Liu, "User participation game in collaborative filtering," in *2014 IEEE Global Conference on Signal and Information Processing (GlobalSIP)*, Dec 2014, pp. 263–267.
31. M. Lichman, "UCI machine learning repository," 2013. [Online]. Available: http://archive.ics. uci.edu/ml

Chapter 7
Conclusion

Abstract In previous chapters we have shown how to apply game theory to deal with the privacy issues in different scenarios. Here in this chapter we make a summarization for the proposed approaches.

Personal data are the fundamental resources of big data applications. The exploration of personal data can create significant value, meanwhile, individuals' privacy will be compromised. It is important to achieve a balance between data exploration and privacy protection. The exploration of personal data generally involve multi-stakeholders. Considering the interactions among the stakeholders, we treat the stakeholders as players of a game, and analyze the game to find equilibrium strategies of the stakeholders. In this book, we present a comprehensive review of our recent research progress on data privacy game. The contribution of our work can be summarized as follows:

We differentiate four different user roles that are commonly involved in data mining applications, i.e. data provider, data collector, data miner and decision maker. For each user role, we discuss its privacy concerns and the methods it can adopt to protect privacy. Based on the user role model, in Chap. 2 we build sequential game model to analyze the following data collecting scenario: a data collector collects data from data providers and then publish the data to a data miner. The data collector performs data anonymization so as to protect data providers' privacy. However, anonymization causes a decline of data utility. Consequently, the data miner will suffer a loss. We apply backward induction to find the subgame perfect Nash equilibria of the proposed sequential game. Simulation results show that the game theoretic analysis can provide guidance to both the data collector and data miner on the trade-off between data providers' privacy and data utility.

In the game model proposed in Chap. 2, we ignore the differences among data providers' privacy preference and consider all data providers as a whole. In Chap. 3, we study the interactions between the data collector and data providers by considering each data provider individually. Different data providers treat privacy differently, and their privacy preferences are unknown to the collector. That is, there is information asymmetry between the collector and providers. We propose a contract theoretic approach for data collector to deal with the data providers.

By designing an optimal contract, the collector can make rational decisions on how to pay incentives to data providers and how to adjust the parameter(s) of the anonymization algorithm so as to protect data providers' privacy. Specifically, we treat the privacy protection level realized by anonymization as a contract item, and explicitly solved the optimal production functions and information rent functions for any given protection level. We've shown that as the collector's requirement on data changes, the optimal functions may be formed in a different way. As for the optimal privacy protection level, we've analyzed how it should be adjusted when the collector faces a different requirement on data utility or has a new valuation of data. Such analysis can provide a practical guidance for private data collecting.

The optimal contract proposed in Chap. 3 bases on an assumption that data providers' privacy preferences are randomly drawn from a distribution which is known to the data collector. In Chap. 4, we relax this assumption a bit. That is, we still assume that data providers' privacy preferences are randomly drawn from a distribution. However, this distribution is unknown to the collector. We consider a scenario where a data collector sequentially buys data from multiple data providers. A data provider's privacy preference is indicated by his valuation of the data. Each time a new data provider arrives, the collector offers the provider a price, and the provider will sell his data if and only if the price is higher than his valuation of the data. To maximize the total payoff, the collector needs to dynamically adjust the prices offered to providers. We model the pricing problem as a multi-armed bandit problem. Specifically, the data anonymization technique adopted by the collector is taken into account. Due to the information loss caused by anonymization, the distributions of rewards associated to the arms are time-variant. Based on the classic upper confidence bound policy, we propose several learning policies to adapt to the time variant characteristic. Simulation results demonstrate that the proposed learning polices can bring the collector a good payoff. And based on the learning results, the collector can make the best decision if he needs to set a single price for data providers.

Previous chapters mainly focus on the strategies of the data collector. In Chaps. 5 and 6, we study how the owners of data should behave when providing data to others. In Chap. 5, we build a game model to analyze users' rating behaviors in a collaborative filtering-based recommendation system. The set of items rated by a user is seen as the user's strategy. A user can get high-quality recommendations only when both the user himself and other users providing sufficient rating data to the recommendation server. However, providing more ratings generally implies disclosing more privacy. We assume that each user has an expectation for the recommendation quality, thus an ideal outcome of the game is that every user has got satisfying recommendations. We propose an algorithm to learn the satisfaction equilibrium of the game. The learning algorithm basically defines a behavior rule which allows the user iteratively updates the probability distribution over his action space and gradually rate more items. We have demonstrated via simulations that when users have moderate expectations for recommendation quality and satisfied users are willing to provide more ratings, then all users can get satisfying recommendations without providing many ratings.

In Chap. 6, we model the interactions among users in a distributed classification scenario as a game in satisfaction form. We assume that there is an untrustworthy mediator who collects data from users and trains the classifier. The mediator sends count queries to users, and users add noise to the results so as to meet the differential privacy criterion. The accuracy of the classifier is affected by the noise which is determined by the privacy budgets chosen by users. We treat the privacy budget as the user's strategy. Similar as before, we assume that each user has an expectation for the classification accuracy. And a learning algorithm is proposed to find the satisfaction equilibrium of the game. By conducting simulations on real-world data, we have demonstrated that when users have similar expectations for the accuracy, the proposed learning algorithm can converge to a satisfaction equilibrium, at which every user achieves a balance between the classification accuracy and the preserved privacy.

We believe that above studies can demonstrate that game theory is well suited for modeling privacy-related scenarios. We hope that the work introduced in this book can offer researchers a new insight into the privacy issue, and promote the exploration of interdisciplinary solutions to achieve the balance between data exploration and privacy protection.

Printed in the United States
By Bookmasters